TOOL-S

A CONCISE F

ON

TOOL-STEEL IN GENERAL,

ITS TREATMENT IN THE OPERATIONS OF FORGING, ANNEALING,
HARDENING, TEMPERING, ETC.

AND THE APPLIANCES THEREFOR.

BY

OTTO THALLNER,

IRON-MASTER AND MANAGER IN CHIEF OF THE TOOL-STEEL WORKS,
BISMARCKHUTTE ON THE SAALE, GERMANY.

AUTHORIZED TRANSLATION FROM THE GERMAN,

BY

WILLIAM T. BRANNT.

British Library Cataloguing-in-Publication Data
A catalogue record for this book is available from the
British Library

Metal Work

Metalworking is the process of working with metals to create individual parts, assemblies, or large-scale structures. The term covers a wide range of work from large ships and bridges to precise engine parts and delicate jewellery. It therefore includes a correspondingly wide range of skills, processes, and tools. The oldest archaeological evidence of copper mining and working was the discovery of a copper pendant in northern Iraq from 8,700 BC, and the oldest gold artefacts in the world come from the Bulgarian Varna Necropolis and date from 4450BC. As time progressed, metal objects became more common, and ever more complex. The need to further acquire and work metals grew in importance. Fates and economies of entire civilizations were greatly affected by the availability of metals and metalsmiths. The metalworker depends on the extraction of precious metals to make jewellery, buildings, electronics and industrial applications, such as shipping containers, rail, and air transport. Without metals, goods and services would cease to move around the globe with the speed and scale we know today.

One of the more common types of metal worker, is an iron worker – who erect (or even dismantle) the structural steel framework of pre-engineered metal buildings. This can even stretch to gigantic stadiums and arenas, hospitals, towers, wind turbines and bridges. Historically ironworkers mainly worked with wrought iron, but today they utilize many different materials including ferrous and non-ferrous metals, plastics, glass, concrete and composites. Ironworkers also unload, place and tie reinforcing steel bars (rebar) as well as install post-tensioning systems, both of which give strength to the concrete used in piers, footings, slabs, buildings and bridges. Such labourers are also likely to finish buildings by erecting curtain wall and window wall systems, pre-cast concrete and stone, stairs and handrails, metal doors, sheeting and elevator fronts – performing any maintenance necessary.

During the early twentieth century, steel buildings really gained in popularity. Their use became more widespread during the Second World War and significantly expanded after the war when steel became more available. This construction method has been widely accepted, in part due to cost efficiency, yet also because of the vast range of application – expanded with improved materials and computer-aided design. The main advantages of steel over wood, are that steel is a 'green' product, structurally sound and manufactured to strict specifications and tolerances, and 100% recyclable. Steel also does not warp, buckle, twist or bend, and is therefore easy to modify and maintain, as well as offering design flexibility. Whilst these advantages are substantial, from aesthetic as well as financial points of view, there are some down-sides to steel construction. It conducts heat 310 times more efficiently than wood, and faulty aspects of the design process can lead to the corrosion of the iron and steel components – a costly problem.

Sheet metal, often used to cover buildings in such processes, is metal formed by an industrial process into thin, flat pieces. It is one of the fundamental forms used in metalworking and it can be cut and bent into a variety of shapes. Countless everyday objects are constructed with sheet metal, including bikes, lampshades, kitchen utensils, car and aeroplane bodies and all manner of industrial / architectural items. The thickness of sheet metal is commonly specified by a traditional, non-linear measure known as its gauge; the larger the gauge number, the thinner the metal. Commonly used steel sheet metal ranges from 30 gauge to about 8 gauge. There are many different metals that can be made into sheet metal, such as aluminium, brass, copper, steel, tin, nickel and titanium, with silver, gold and platinum retaining their importance for decorative uses. Historically, an important use of sheet metal was in plate armour worn by cavalry, and sheet metal continues to have many ornamental uses, including in horse tack. Sheet metal workers are also known as 'tin bashers' (or 'tin knockers'), a name derived from the hammering of panel seams when installing tin roofs.

There are many different forming processes for this type of metal, including 'bending' (a manufacturing process that produces a V-shape, U-shape, or channel shape along a straight axis in ductile materials), 'decambering' (a process of removing camber, or horizontal bend, from strip shaped materials), 'spinning' (where a disc or tube of metal is rotated at high speed and formed into an axially symmetric part) and 'hydroforming.' This latter technique is one of the most commonly used industrial methods; a cost-effective method of shaping metals into lightweight, structurally stiff and strong pieces. One of the largest applications of hydroforming is in the automotive industry, which makes use of the complex shapes possible, to produce stronger, lighter, and more rigid body-work, especially with regards to the high-end sports car industry.

One of the most important, and widely incorporating roles in metalwork, comes with the welding of all this steel, iron and sheet metal together. 'Welders' have a range of options to accomplish such welds, including forge welding (where the metals are heated to an intense yellow or white colour) or more modern methods such as arc welding (which uses a welding power supply to create an electric arc between an electrode and the base material to melt the metals at the welding point). Any foreign material in the weld, such as the oxides or 'scale' that typically form in the fire, can weaken it and potentially cause it to fail. Thus the mating surfaces to be joined must be kept clean. To this end a welder will make sure the fire is a reducing fire: a fire where at the heart there is a great deal of heat and very little oxygen. The expert will also carefully shape the mating faces so that as they are brought together foreign material is squeezed out as the metal is joined. Without the proper precautions, welding and metalwork more generally can be a dangerous and unhealthy practice, and therefore only the most skilled practitioners are usually employed.

As is evident from this incredibly brief introduction, metalwork, and metalworkers more broadly, have been, and still are – integral to society as we know it. Most of our modern buildings are constructed using metal. The boats, aeroplanes, ships, trains and bikes that we travel on are constructed via metalwork, and mining, metal forming and welding have provided jobs for thousands of workers. It is a tough, often dangerous, but incredibly important field. We hope the reader enjoys this book.

PREFACE.

THE changes which tool-steel undergoes by the various operations of forging, annealing, hardening, tempering, etc., have theoretically been established by Ledebur, Wedding, Reiser, Osmond, and others.

The rules, which have been deduced from theory, of what has to be observed in the above-mentioned operations are in themselves of a simple nature and readily comprehended, and a more universal knowledge of them has been diffused by the scientific publications of the above-mentioned writers, as well as by the so-called " directions for the treatment" of the steel, which nearly every manufacturer of tool-steel furnishes to his customers.

The almost sole object of these directions is to promote a more intimate knowledge of steel and its treatment in the manufacture of tools among those who are especially entrusted with it, but, as a rule, an explanation of "how and by what means" their observance is to be effected is wanting.

The steel recommended by them is not seldom invested with mysterious properties, and encouragement is frequently given to the continuance of primitive appliances, but little suitable for the purpose and difficult to attend.

In these directions the pith of the matter is very seldom treated of, or only in a very brief way, so that the changes which the steel undergoes in the various operations of the manufacture of tools remain, as a rule, not understood, and there is no probability of the mind being directed towards the necessity of suitable working appliances.

Publications on this subject derived from practice are also scarce, because experiences gathered in this line are preferably kept secret.

These conditions apparently explain the limited general diffusion of a knowledge of practically approved appliances and working processes, in consequence of which master-workman and manufacturer are frequently placed in the disagreeable position of having to desist from making the most of the highest efficiencies attainable in tools, or are forced to work with a greater expense of time, money and material than would otherwise be necessary.

The master-workman and the workmen entrusted with the manufacture of tools are under the necessity of gathering experience from their own practice, and it is largely left to themselves to devise appliances required for successful working without having at their disposal anything that will give them a clue to *how* it is to be done.

It was especially the latter condition that induced me to write this small work.

It is chiefly intended as a guide to the master-workman and the intelligent tool-maker, and, in accordance with this object, is exclusively adapted to practical needs.

As sources of the explanations derived from theory, which have been incorporated into the book, the scientific works and separate publications of the previously mentioned authors have served.

The general arrangement of the material has been modeled after the excellent work by F. Reiser: "Das Härten des Stahles in Theorie und Praxis" ("Hardening of Steel in Theory and Practice").

The directions and working appliances collected in the book have throughout been taken from practice, and are intended to assist master-workmen and workmen employed in the practical execution of the various operations in the manufacture of tools, in their occupation, which demands varied knowledge and experience.

THE AUTHOR.

BISMARCKHUETTE ON SAALE.

CONTENTS.

INTRODUCTION.

I.

COMPOSITION OF TOOL-STEEL AND ITS CLASSIFICATION ACCORDING TO IT.

II.

CLASSIFICATION OF TOOL-STEEL ACCORDING TO THE DEGREE OF HARDNESS AND THE PURPOSE FOR WHICH IT IS TO BE USED.

III.

OBSERVATIONS ON THE EXTERNAL APPEARANCE OF COMMERCIAL TOOL-STEEL.

IV.

OBSERVATIONS ON THE FRACTURE OF STEEL, WITH REGARD TO THE STRUCTURE IN THE HARDENED AND NON-HARDENED STATES.

V.

PRACTICE OF THE FIRE-TREATMENT OF STEEL.

VI.

APPLIANCES FOR ANNEALING STEEL.

VII.

APPLIANCES FOR HARDENING STEEL.

VIII.

HARDENING OF TOOL-STEEL IN GENERAL.

IX.

HARDENING OF TOOLS WHICH ARE TO BE HARDENED IN THEIR ENTIRETY.

X.

HARDENING OF TOOLS WHICH ARE ONLY TO BE PARTIALLY HARDENED.

XV.

CASE-HARDENING AND PREVENTATIVES AGAINST SUPERFICIAL DECARBONI-ZATION AND OVERHEATING.

XVI.

WELDING OF STEEL.

XVII.

REGENERATION OF STEEL WHICH HAS BEEN SPOILED IN THE FIRE.

XVIII.

INVESTIGATIONS OF DEFECTS OF HARDENED TOOLS.

XIX.

IMPROVING THE PROPERTIES OF STRENGTH OF STEEL.

APPENDIX.

TOOL-STEEL.

INTRODUCTION.

In a wider sense, nearly all the products of iron-works, commencing with pig iron up to weld-iron and ingot-iron, which are not capable of being hardened, are employed in the manufacture of tools. However, in a narrower sense, iron which is capable of being hardened, *i. e.*, steel, has, under the general designation of *tool-steel*, to be chiefly taken into consideration for the manufacture of tools.

In many cases tool-steel is still more closely designated according to the smelting processes from which it has resulted, such as ingot-steel, weld-steel; or according to the method of production, such as Bessemer steel, Martin steel, crucible steel, puddled steel, shear steel, etc.; further according to certain peculiarities, such as natural or self-hardened steel, hard-centred steel, soft-centred or mild-centred steel, tungsten-steel, chrome-steel, nickel-steel, etc.; and finally, according to the purposes to which it is to be applied, such as knife-steel, scythe-steel, magnet-steel, punch-steel, chisel-steel, etc.

Beside the above-mentioned designations, there are also in use several of an allegorical nature, such as diamond-steel, self-hardener, boreas, atlas-steel, universal steel, etc., and others which intimate an alloy—which, however, as a rule, cannot be authenticated—to improve the quality of

(1)

the steel, such as titanium-steel, aluminium-steel, vanadium-steel, etc.

The most noble product of the manufacture of tool-steel is *crucible-steel, i. e.*, steel made by remelting in a crucible basis-materials in themselves very pure and of excellent quality. The object of this process is to obtain a thoroughly uniform product as free as possible from injurious admixtures, and in the preparation of it every precaution is taken to avoid mistakes which might impair the quality of the finished tool-steel. Hence crucible-steel is rather expensive and for this reason ingot-steel produced by the Bessemer and Martin processes is largely used for inferior purposes.

This steel is generally brought into commerce under the simple designation of " cast steel," and by many consumers is worked in good faith as crucible steel.

The steel for the manufacture of tools brought into commerce under the name of " weld-steel " is practically, as a rule, steel capable of being welded or hardened without regard as to whether or not it is weld-steel according to the smelting processes by means of which it has been produced. In most cases the product brought into commerce under the name of weld-steel is " weldable ingot steel."

I.

COMPOSITION OF TOOL–STEEL AND ITS CLASSIFICATION ACCORDING TO IT.

Iron, which by being heated to a red heat and suddenly cooled, has acquired such hardness that it is not attacked by a file is, generally speaking, called *steel*. This property of iron results from a content of carbon which may amount to between 0.5 and 2 per cent.

Iron with less than 0.5 per cent. carbon * does not acquire such a high degree of hardness, and with more than 2 per cent. carbon, it becomes pig-iron, and beside losing the capacity of being hardened, also loses the property of malleability.

It has been shown by scientific researches that in iron, carbon occurs in various forms, the most important of which are as follows:

  a. Carbon fixed to the iron forming a chemical combination with it (hardening carbon).

This form of carbon appears to be the carrier of the hardening capacity in steel, and the process which takes place thereby has scientifically been determined as follows:

When iron containing carbon is allowed slowly to cool

---

* The transition from iron incapable of being hardened to steel which can be hardened is a gradual one, and the limit between them cannot be expressed by the content of carbon alone. The limit mentioned above refers to crucible-steel of the lowest degree of hardness used for hardened tools.

from the melting temperature, the carbon present forms with a portion of the iron a combination which is net-like diffused throughout the iron.   (Carbide.)

When such iron is heated to a cherry-red heat, the carbon present is uniformly absorbed (dissolved) by the iron and becomes hardening carbon.   By quickly cooling the heated steel the carbon becomes fixed in this state, and the steel is hardened.   If, however, the steel is allowed to cool slowly, the carbon is separated from its uniform combination with the iron and again forms with a portion of the latter the net-like diffused carbide.

The larger the quantity of carbon (carbide) present, the more of it can during the process of hardening be converted into hardening carbon, and the greater will be the degree of hardness which the steel acquires.

b. The carbon may occur in a free form visibly imbedded between the particles of iron (graphite).

Graphitic carbon cannot be converted by the operation of hardening into hardening carbon, and hence does not induce hardening.   It is a chief constituent of gray pig iron. It occurs very seldom in tool-steel and, when present, exerts an injurious influence upon its quality.

Besides iron and carbon, steel contains other substances which are either accidental admixtures, or are added for a special purpose.   Phosphorus, sulphur, copper and arsenic are the principal accidental admixtures, and they exert always an injurious effect.   A few hundredths of one per cent. of either of these substances suffice to render steel partially or entirely unfit for the manufacture of tools. The quality of steel is dependent on the quantity of injurious constituents it contains, and the total sum of them may serve as an expression in figures of its value.

In tool-steel it is sought to avoid, as much as possible, the occurrence of injurious impurities, and in its production materials free from them are, as a rule, only used. Hence, the quantity of injurious admixtures in tool-steel is never so large that their presence could at once be inferred from its physical behavior in working.

No matter what the nature of the injurious admixtures may be, it shows itself chiefly by brittleness in tool-steel when in the hardened state, and with an abundance of impurities, also in the unhardened state.*

Crucible cast-steel generally contains silicon as a fortuitous admixture which has been absorbed by melting in the crucible. This content of silicon is, however, seldom so great as to exert an essential influence upon the quality of the steel. A higher content of silicon promotes the edge-holding capacity, but also increases the brittleness of the steel, and is injurious when it causes a partial separation of the carbon in graphitic form. The fracture of such steel shows a dark grain.

Intentionally are added to tool-steel the following metals, with the expectation of improving thereby the quality:

* Numerous chemical and practical investigations in Bismarckhütte regarding the qualitative value of tool-steel dependent on the degree of impurities have shown that the total amount of phosphorus, sulphur and copper should not exceed 0.06 per cent., if the material is, according to practical conditions, to be classed as "very good" tool-steel. Steel which contains a total of 0.10 per cent. of phosphorus, sulphur and copper may be designated as "good" tool-steel, and when it contains more than this amount, as "medium" to "bad" tool-steel.

Since the tenacity of steel decreases with an increase in the content of carbon, the effect of injurious admixtures is essentially greater in harder than in softer steel, so that the figures given above may be somewhat raised with soft steel and somewhat lowered with hard steel.

Manganese, tungsten, chromium, nickel, and more seldom molybdenum and other metals.

MANGANESE.

Every kind of tool-steel contains manganese, the quantity in the ordinary product being from 0.2 to 0.5 per cent. Within these narrow limits manganese does not exert any considerable influence upon the properties of the steel, the strength, hardness and cutting power being slightly increased. On the other hand, it is the task of manganese to remove or fix during the fusing processes the gases and the oxides which are present in great abundance in liquid steel, and thus cause the production of dense castings free from blisters.

Under the name "manganese steel," a tool-steel with the normal content of manganese is brought into commerce. Very likely it bears this name only for the purpose of indicating that the steel has been melted with a content of manganese, and hence is freer from pores and cracks than other steel. *Actual manganese steel* which contains a considerable quantity of manganese (8 to 20 per cent.) is particularly strong, tough, and of such great natural hardness that it can scarcely be worked.

Such manganese steel is brought into commerce in the form of finished tools, such as hatchets, axes, etc., or as material for machine-parts which are to possess special hardness and strength.

TUNGSTEN.

Tungsten is added to tool-steel if its hardness is to be essentially increased and its properties of strength are to be improved. Up to 10 per cent. of tungsten is found in tool-steel; seldom more, but generally 2 to 4 per cent., and occasionally—especially in English steel—less than 1 per cent.

A larger content of tungsten—over 2 per cent.—imparts to the steel, even while in an unhardened condition, a fine-grained structure of characteristic lustre Hardened steel alloyed with tungsten shows even in the presence of very small quantities of it a very fine grain of dull lustre. With a higher content of tungsten the structure becomes so fine and velvety that the grain can scarcely be recognized with the naked eye.

By a content of tungsten the hardness and edge-holding power of the steel are increased, but in a hardened state its tenacity is decreased. Hence, tool-steel with a content of tungsten is generally used only for tools which are to be gently engaged and are to possess great hardness and edge-holding power. By repeated treatment in the fire, steel which contains tungsten or chromium loses its good qualities much more rapidly than steel free from these constituents, the edge-holding power decreasing very rapidly, while brittleness and tendency towards cracking in hardening increase. The sensitiveness of such steel towards overheating is also much greater.

CHROMIUM.

Chromium is added to tool-steel for the same purpose as tungsten, but not with the same result, since it does not possess the same efficiency and makes the steel much more brittle.

Up to 3 per cent. of chromium is found in tool-steel; generally 2.5 per cent. in very hard turning tools, and more seldom less than 1 per cent. in softer steels (in some English varieties of steel).

The property of chromium to make steel especially resistant to blow and shock has led to the special use of chrome-steel for shells, and in conjunction with nickel, for armor plates.

The appearance of the fracture of steel is influenced by a content of chromium in a similar manner as by a content of tungsten.

NICKEL.

Nickel greatly improves the strength of steel, its hardness and tenacity in an unhardened state being essentially increased. In a hardened state nickel does not act in the same degree as tungsten or chromium, so that its use in tool-steel may be left out of consideration and nickel-steel to be employed in an unhardened state need chiefly be mentioned.

Machine-parts on which heavy demands are made, and which, either to save weight or space, are to have as small dimensions as possible, or the strength of which is to be especially increased, such as propeller-shafts, crank-pins, etc., are frequently constructed of nickel-steel, but its greatest application is for armor-plate, cannon, etc.

Nickel-steel as a rule contains 6 to 7 per cent. nickel. With a content of over 5 per cent. the steel possesses the most favorable properties of strength, but is also of such natural hardness that it can scarcely be worked in the cold state.

MOLYBDENUM, TITANIUM, VANADIUM.

Molybdenum is seldom alloyed with tool steel because

the high price of the metal does not allow of its extensive use, and as its action is very similar to that of tungsten the same result can be attained with the latter metal.

Titanium and vanadium are not employed in the manufacture of tool-steel on account of their high price and the difficulty in producing varieties of steel alloyed with them.

Titanium and vanadium steels have been made on a small scale for the purpose of testing their properties, and titanium has been found to impart special hardness, and vanadium special tenacity, to steel.

Nevertheless varieties of steel—especially of English origin—are brought into commerce under the names of molybdenum-steel, titanium-steel, vanadium-steel, but of course they do not contain any of these constituents.

II.

CLASSIFICATION OF TOOL-STEEL ACCORDING TO THE DEGREE OF HARDNESS AND THE PURPOSE FOR WHICH IT IS TO BE USED.

As regards the varieties of tool-steel brought into commerce two groups may be distinguished so far as products of crucible steel are concerned, namely :

a. Tool-steel which acquires its hardness exclusively from a content of carbon and does not contain any admixtures which increase the hardness.

b. Tool-steel, which in addition to carbon, contains admixtures increasing the hardness. In practice it is generally called special steel, or after the kind of admixture—chrome-steel, tungsten-steel, nickel-steel, etc.

The tool-steel brought into commerce is nearly always provided with the stamp of the firm and a colored printed label. The latter, in addition to the name of the firm and trade-mark, contains data regarding the degree of hardness, the principal purposes for which the steel is to be used, and the temperature to be employed in forging and hardening.

In some manufactories of tool-steel the degree of hardness is expressed by figures corresponding to the content of carbon, for instance, degree of hardness 7—containing 0.70 per cent. carbon—while in others it is expressed in such allegorical designations as: very hard, super-hard, extra hard, hard-hard, medium hard, tenaciously hard, tenacious, very tenacious, soft. In addition to these designations the content of carbon in per cent. is occasionally given.

The color of the paper label is generally so selected that the hardest steel is provided with a label of the lightest, and the softest steel with one of the darkest, color.

The varieties designated special steel, which as a rule are alloyed with tungsten or chromium, are in practice provided with labels of different text and colors from those of ordinary tool-steel.

While the above-mentioned allegorical designations for the degree of hardness of tool-steel are in use in most factories, not all of them understand under the same designation the same practical degree of hardness according to the content of carbon. Thus one factory designates steel with 1.2 per cent. of carbon as super-hard, while another terms it medium hard or hard, and steel with 0.6 per cent. carbon is designated as soft, tenacious and even tenaciously hard.

Nevertheless in practice, these designations serve their purpose, their chief object being to enable the consumer to procure steel of the same degree of hardness from the same source of supply. In judging the fitness of a quality of steel for a definite purpose the percentage of carbon alone is by no means decisive.

It may be accepted as a general rule that for a determined purpose a tool-steel may be selected which may be the harder (richer in carbon), the freer from injurious admixtures, sulphur, copper, phosphorus, etc., it is.

The higher the percentage of carbon in a steel which for a determined purpose may be chosen, the greater will be the useful effect which may be expected of a tool made therefrom, but the more care must also be bestowed upon the treatment of the tool while being manufactured.

However, in practice, the useful effect of the finished tool is not always taken as the standard for judging the fitness of a tool-steel. In the majority of cases steel is judged according to the manner in which it can be worked into finished tools without making special demands on the intimate knowledge and attention of the tool-smith, forgeman or hardener, and without the requirement of special devices for the most important operations in the manufacture of tools, namely, forging, annealing, hardening, tempering, etc.

Soft steel is less exposed to the danger of over-heating and burning than hard tool-steel, and by reason of its greater toughness, is less liable to crack in hardening. Hence it requires less attention in working and consequently is more largely used in the practice than hard steel. This is one of the reasons for the extensive employ-

ment of soft Bessemer and Martin steels in the manufacture
of tools, even where the advantage of the greater useful
effect of crucible cast-steel, which is used much harder,
should be appreciated. To be sure this greater useful effect
is not alone attained by the employment of harder steel,
very conscientious and careful work in the manufacture
of the tools being also required, as well as thoroughly-
informed and experienced forge-men and hardeners.

From what has been said in the foregoing, it seems useless to lay down a rule which would be even of general
value for the degree of hardness to be chosen for determined purposes. For this reason the classification of steel
according to the degree of hardness and the purposes for
which it is to be employed, as in use in Bismarckhuette,
and which on an average has proved of practical value,
will here only be given as follows:

Degree of hardness.	Average per cent. of carbon.	Purposes for which Employed.
Very hard.......	1.5	For turning and planing knives, drills, turning gravers, etc., for very hard materials.
Hard	1.25	For ordinary turning and planing knives, rock drills, mill picks, knife picks, scrapers, etc., and for cutting tools for hard metals.
Medium hard....	1.0	For screw-taps, broaches, cutters, tools for stamping presses and for various tools used by locksmiths and blacksmiths.
Tenaciously hard.	0.85	For screw-taps, cutters, broaches, matrices, swages, pins, bearings, chisels, gouges, etc.
Tough	0.75	For chisels and gouges, shear-blades, drifts, springs, hammers, etc.
Soft	0.65	For various blacksmith tools, as weld-steel for steeling finer tools and larger surfaces, etc.

SPECIAL STEELS.

The principal commercial varieties of special steel, the

hardness of which has, in addition to carbon, been increased by a content of tungsten or chromium, are as follows:

Natural tool-steel, also called self-hardening, boreas, mushet steel.

Special turning steel.

Magnet steel.

Natural or self-hardened steel in an unhardened state possesses such great cutting power and hardness that in this state it can be employed for cutting tools which are not subjected to shock.

When slowly cooled from a red heat the hardness of natural steel is greater than when rapidly cooled from this temperature, and its behavior in hardening is thus the reverse from ordinary tool-steel.

Self-hardened steel has the advantage of retaining its hardness when heated, and is therefore suitable for lathe-cutters upon hard materials in lathes running at high speed, whereby heat is generated, or for taking off a chip of extra thickness, etc. Such cutters keep their edges much better than cutters of ordinary steel. This property of self-hardened steel is due to a higher content of tungsten, manganese and silicon. The composition of self-hardened steel is shown by the analyses given below:

	Carbon.	Manganese.	Silicon.	Tungsten.
English mushet steel..........	1.71	1.8	0.81	7.75
Styrian steel	1.78	1.85	1.01	9.72
Bismarckhütte natural steel ...	2.04	1.78	1.08	9.50

Very hard special turning steel contains, in addition to 1 to 1.5 per cent. carbon, 3 to 6 per cent. tungsten, manganese and silicon in the usual quantities found in other tool-steel. It contains very seldom a higher percentage of manganese or silicon with a smaller content of tungsten, as, for instance, the special steel produced by Marsh Bros., which with only 1.8 per cent. of tungsten contains 1.8 per cent. of manganese.

All varieties of special steel, the hardness of which is considerably increased by a larger content of tungsten, require specially careful treatment in hardening if their hardness is to be thoroughly effective without the tool succumbing to the correspondingly enormous hardening strains and cracking. To insure suitable treatment in hardening, such varieties of special steel are generally accompanied by special, explicit instructions.

Magnet steel * has a composition similar to that of special steel, and generally contains as large a percentage of tungsten, the latter exerting considerable influence upon the improvement of the magnetic properties of the steel.

Since the cutting power of magnet steel is of secondary importance, it is generally given such a chemical composition as to make its magnetic properties most effective.

There are numerous varieties of special steel, the chemical composition of which does not differ from that of ordinary steel. They are generally designated by names indicative of the purpose they are to serve, and for which they are claimed to afford the greatest efficiency attainable.

Tool-steel for definite purposes, the cross section of which

---

* It may be accepted as a general rule that the coercive power, *i. e.*, the power with which the magnetism is retained, and the quantity of magnetism taken up, are the greater the harder the steel is. The admixtures of manganese and silicon in steel, which can never be entirely avoided, exert an influence according to the quantities present.

shows different degrees of hardness, is generally produced
by welding together iron or mild steel with hard steel in
casting the crude block. The accompanying sketches, Fig.
1, show such varieties of steel, the hatched lines represent-
ing hard steel.

Fig. 1.

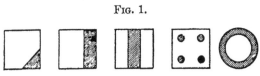

Such steel is used for a variety of cutting tools, the edges
of which consist of hard steel and the backing generally of
iron ; further for safes, calks for horse shoes, etc.

The variety of tool-steel brought into commerce under
the name of soft-centred or mild-centred steel is hardest on
the surface, while the interior is very soft. The transition
from hard surface to soft interior is a gradual one and
seldom sharply defined. Such steel is produced by cementation of mild steel, carbonization being arrested at a
definite stage. This mode of production frequently results
in an unequal material, and for this reason such steel is
seldom used in practice for the manufacture of tools. It
is, however, employed to advantage for machine-parts which
have to be very tough, but portions of the surfaces of which
have to be harder to protect them against rapid wear, for
instance, crank pins, dynamo shafts, etc., the bearing portions of which are hardened while their interior are left
tough and soft.

III.

OBSERVATIONS ON THE EXTERNAL APPEARANCE OF COMMERCIAL TOOL–STEEL.

When receiving tool-steel, the eye of the consumer is involuntarily drawn towards its external condition. So far as the quality of the steel is concerned, the most perceptible features in this respect are the appearance of the fracture and the cleanness of the surface.

With steel carefully made defects can seldom be noticed by the fracture and surface.

The surface of steel may show the following defects:

Scales, the form of which represents a single line running in a curve, or the line may consist of several coherent curves. Such scales generally originate from dross on the surface of the block which has not been removed, from remnants of slag or from skins formed in casting the steel. They are more seldom due to faults in forging or rolling.

Cracks are outlined on the surface of the steel by short lines, either singly or in groups, running parallel to the longitude of the surface. They are partially concealed from view by the layer of oxide covering the steel, but may readily be exposed by means of a file. These cracks are due to pores which in casting the steel-ingot have been formed immediately below the surface.

Seams are formed by the steel protruding laterally in rolling or forging in the swage and by turning down and squeezing in the protruded material in the further working of it. The seams run always parallel to the longitude of the steel, generally on two sides or edges of it; more seldom alternately, or on one side only. On rolled steel such seams readily escape observation when they are covered with a layer of oxide.

Edge-cracks are cross-rents on the edges which run vertically to the direction of the length of the steel. They indicate red-shortness or hot-shortness, or that the steel in forging has been strongly overheated (burnt). Such steel is of course entirely unfit for the manufacture of tools.

In the further working of the steel the principal defects noticeable on the surface, which have been mentioned above, are frequently productive of defective tools because cracks formed in hardening originate from them.

The appearance of the fracture is not decisive as regards the quality of unhardened steel, and only within very uncertain limits as regards its hardness.

The structure of soft steel shows a coarse grain, and that of hard steel generally a fine grain, the latter being the finer the more carbon the steel contains, provided the surfaces of the fractures which are to be compared have been produced under the same conditions.

If the fracture of very soft steel shows a proportionately coarse-grained structure and at the same time a dark streak towards the edge in which an actual grain can no longer be recognized, but an almost fibrous structure, it is indicative of very soft steel with a very small content of carbon —below 0,6 per cent., with less than 0.3 per cent. manganese.

If a dark, lustreless streak is noticeable on the fracture of hard steel which otherwise shows a fine-grained structure, it is indicative of the presence of carbon in graphitic form.

2

Such steel acquires only an unequal, insufficient degree of hardness and always yields bad tools.

The following defects, if present, may be observed on the surfaces of the fractures of tool-steel:

1. *Flaws or blisters*, in the center, and more seldom, towards the edge of the steel.

   When towards the edge of the steel, they are due to pores, and when in the centre of the fracture, to the pipe.

2. A spot, generally symmetrically formed, in the centre of the steel, which shows a coarser or finer structure than the surrounding parts. Such spots are formed by liquation on cooling steel freshly cast, and their chemical composition differs from that of the surrounding portions. When the fracture of such steel is filed, ground, polished and pickled in acid, spots of unequal hardness (unequal chemical composition) are outlined as quite sharply defined surfaces of different coloration.

3. When cutting up a bar of steel into the separate parts serving for the manufacture of a tool, flaws may also be noticed in the centre of the steel. Such flaws are only in rare cases attributable to a continuation of the pipe, but are more frequently due to destructive forging. In forging bar-steel it is nearly always worked colder in the centre than on the ends, the interior being readily shattered thereby. The flaws thus formed seldom extend to the surface of the steel, and, therefore, escape detection in the works where the steel is produced. The defect caused by destructive forging is more fre-

quently found in hard steel and in material of small or thin dimensions.

In most tool-steel works it is sought to avoid the above-mentioned defects which can be noticed on the surface and fracture of the steel, and the various products are subjected to very careful inspection before being shipped to prevent defective material from reaching the consumer.

IV.

OBSERVATIONS ON THE FRACTURE OF STEEL, WITH REGARD TO THE STRUCTURE IN THE HARDENED AND NON-HARDENED STATES.

IT requires considerable practice and much experience to be able to judge, with any degree of certainty, from the appearance of an otherwise faultless fracture, the quality and incidentally the hardness of a variety of steel.

As previously mentioned, the structure of steel becomes more fine-grained as the hardness increases. However, steel of the same degree of hardness may show an entirely different structure if before being broken it has been worked in different ways, and after having been worked, has been cooled from high temperatures of different degrees.

Tool-steel which has been heated and then very slowly cooled, and the fracture of which has been effected by nicking and breaking while in the cold state, shows the most coarse-grained structure pertaining to its degree of hardness.

The appearance of the fracture of unhardened steel as brought into commerce is in the main dependent on the following conditions:

1. On the temperature from which the steel has been cooled before it is broken.

   Changes in shape are much more rapidly effected by rolling than by forging, and the steel, when allowed to cool, has a much higher temperature. For this reason rolled steel has, as a rule, a more coarse-grained structure than forged steel.

   The fractures of the two ends of the same bar of forged steel show sometimes a different appearance if, after forging, one-half of the bar has been cooled from a lower or higher temperature than the other.

2. On the degree of mechanical manipulation.

   Tool-steel of the same degree of hardness shows a finer structure in smaller dimensions and a coarser structure in larger dimensions.

   In producing smaller dimensions the loss of heat takes place more rapidly, but the change in shape requires more powerful working of the steel; and this being finally effected at a lower temperature, the fracture shows a fine-grained structure.

3. On the temperature of the steel before it is worked.

   Steel over-heated in heating for forging or rolling acquires an entirely or partially coarse-grained structure, which possesses a characteristic bright lustre different from the normal color of steel.

   This structure frequently forms a border around the edge of the fracture or runs from the edge towards the centre of the steel; and if the steel had been over-heated to a higher degree, does not entirely disappear by subsequent manipulation.

4. On the manner in which fracture has been effected.

Steel which has been nicked in the cold state and then broken shows a more coarse-grained structure than steel which has been nicked in a red-hot state and broken after cooling.

Nicking in a red-hot state is equivalent to a mechanical manipulation at a lower temperature, and hence the finer structure.

The appearance of the fracture of hardened steel is dependent on the following conditions:

1. On the temperature at which the steel by rapid cooling has been hardened.

Suppose a bar of steel has been heated so that the brightest white heat slowly extends from one end to the other, passing through a yellow, red, and brown heat to hand-warm. Now, if this bar is hardened by being quickly quenched in water, the changes in the structure of the steel while being heated and the influence of the various degrees of heating upon the structure, the degree of hardness and the toughness, may be recognized by the separate broken portions and the appearance of their fractures.

The higher the temperature at which hardening has been effected, the more coarse-grained the structure of hardened steel will be.

The accompanying tables show the influence of the various degrees of temperature upon hardened and unhardened steel, and their general application in practice.

2. On the dimensions of hardened steel. Steel with a small cross-section yields up heat more quickly in hardening than steel with a large cross-section. It is more uniformly hardened throughout, and its fracture shows a more uniformly fine-grained structure.

Steel of larger dimensions yields up its interior heat but slowly, and hence does not acquire the same degree of hardness in the interior as towards the surface. The face of the fracture shows in the centre a more coarse-grained structure which becomes more fine-grained towards the edges.

3. The appearance of the fracture in the hardened state is dependent on the structure only when the steel has been overheated or when it has previously been subjected to vigorous mechanical manipulation. In the former case the steel in the hardened state shows a coarser, and in the latter case, an especially fine, structure.

V.

PRACTICE OF THE FIRE–TREATMENT OF STEEL.

In the manufacture of tools repeated heating of the steel is in most cases required, partially for the purpose of shaping (forging), partially for increasing the capacity of being worked (annealing), and for hardening and tempering the finished tool.

Shaping can only in a few cases be effected by one heating, and in many cases a worn-out tool has repeatedly to pass through the above-mentioned operations without its quality suffering injury from repeated heating.

Hence, in the manufacture of tools particular attention has to be paid to every operation for which the steel has to be heated, and the appliances for this purpose should be properly adapted.

This adaptation should be in accordance with the following principles, which must be strictly observed:

1. The steel should always be heated so uniformly that it is not heated more, either in its entirety or partially, than is absolutely necessary for the subsequent operation.

2. Heating should be effected as rapidly as possible without bringing the single parts (corners, edges) of the steel to a higher heat prior to the body.

Though simple and self-evident as these principles may appear, they are frequently transgressed as regards the operation in question itself, as well as in the appliances for carrying it out.

The appliances for heating the steel include in the main fire and furnaces, as well as fuel.

The quantities of fuel and their value are mostly small in comparison with the value of the tool to be produced, and the essentially greater efficiency of a tool which has been carefully made.

The choice of fuel depends partially on local conditions and partially on the manner in which it is to be applied. In many cases the heating devices are adapted to the purpose in view as well as to the fuel at disposal.

The disadvantages which may result, as far as the quality of the steel is concerned, from the use of certain fuels will now be described, as well as the means necessary to successfully avoid them.

Hard coke, which does not stain the hand, and when struck emits a clear sound, is used in the manufacture of tools in the open fire as well as in furnaces. The harder the coke is, the higher the temperature will be, and the more air for the combustion of the coke required.

Hence, in a coke-fire, tool-steel is readily heated too rapidly and too much, and the larger the pieces of coke the more it is exposed to the action of the blast. These are conditions under which the steel may readily become overheated, and even burned.

When no other fuel besides hard coke is available, the furnace shown in Fig. 2 should be used for heating the steel.

For the construction of this furnace take a piece of fire-tube, or similar material, about $3\frac{1}{3}$ feet long and 16 to 24 inches in diameter, and provide it at a, b, c with apertures to which doors of stout sheet iron are fitted. Place the tube thus prepared upon an ordinary brick pavement, line it with fire-brick, as shown in the illustration, fix the blast-pipe w under the grate, and close the upper portion of the brick work with an arch. The cover of the furnace is provided with an aperture $4\frac{3}{4}$ to 6 inches in diameter, in which is fitted a smoke-pipe $3\frac{1}{2}$ to 13 feet high and provided with a damper for regulating the draught.

The fuel is introduced through the doors a, b. The door b is the actual working door, the door a serving for passing through long bars, which are to be heated in the centre for the purpose of dividing them.

The coke burning in the space S heats the working space

A, in which the tool, held by means of tongs or resting
upon a grate-like support, is heated. With a smoke-pipe
of sufficient height, a blast-pipe is not actually required,
since the free draught is sufficient to produce the high and

Fig. 2.

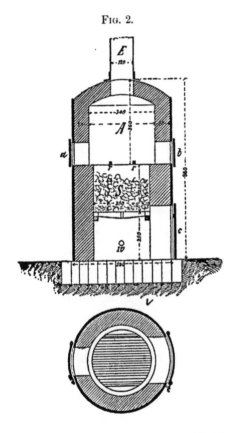

uniform heat required in *A.* The suitable degree of temperature can, within a certain limit, be effected by regulating the draught.

Smith-coal or *forge-coal.* The variety of coal thus designated cakes readily, swelling up thereby, and is most
frequently used for open forge-fires. However, it contains

sometimes a high percentage of sulphur, the presence of which may spoil tool-steel heated in such fire. To remove a content of sulphur, allow the coal to burn thoroughly through until no more smoke is evolved, repeating the operation every time fresh fuel is added.

The effect produced by a content of sulphur in the coal upon the steel heated therein is as follows:

At a high temperature sulphur has a great tendency towards combining with iron. This combination cannot be hardened and causes the formation of so-called soft spots in the steel, i. e., spots of greater or smaller extent on the surface of the steel, which acquire no hardness whatever in hardening the tool.

When smith-coal in an open fire is ignited it readily cakes on the surface to a coherent cover underneath which combustion takes place at a very high temperature. If this cover is not broken up such fire may burn "hollow." If steel for the purpose of heating is brought into the hollow space, it is heated in immediate contact with the blast and is "burnt" at a temperature which, with the exclusion of the oxygen of the air, would not be high enough to spoil it. In heating tool-steel in an open fire it is, therefore, necessary to prevent the coal from forming a solid cover. The development of a less intense heat in a fire frequently loosened is rather an advantage in heating steel.

Coke dust. Coke in minute fragments or reduced to dust is not suitable for use in an open fire, because the blast cannot with sufficient ease penetrate between the interspaces of the fuel to induce sufficiently extensive combustion. Such coke may be used mixed with smith-coal, it readily caking together with it.

The highest degree of heat developed in an open fire is immediately in front of the tuyere, and in the zone surrounding it, within which the blast, scattered by striking the coal and weakened in force, just maintains the coal in full glow, is found the degree of heat required for forging and hardening. This heat, however, decreases too rapidly in intensity towards the surface of the fuel put on.

The heat developed in the middle zone is, of course, not uniform, it being more intense towards the centre of the fire and less so towards the surface. For this reason it will be difficult to heat articles to a uniform heat in this zone when it is narrow, i. e., when but a small quantity of fuel has been put on. The more fuel is put on, the greater will be the width of the heating zone which can be utilized for forging and hardening, and the larger in size the articles may be which can be heated in it. It is extremely difficult to heat in an open fire to a uniform temperature pieces of especially large size, such as anvils, hammer-blocks, swages, etc., it being almost impossible to avoid roasting the steel for several hours. Generally it is also partially over-heated, while some portions are insufficiently heated. In such cases the operation may be essentially facilitated by allowing the blast to enter through two tuyeres arranged parallel at a distance of 80 to 200 inches from each other. The tool imbedded in the space between the nozzles then receives an adequate supply of heat, and needs to be turned less frequently.

When long, thin articles, for instance screw-augers, broaches, shear blades, cutters, etc., are to be heated to a uniform temperature, even skilled forgemen encounter many difficulties on account of the heat not being uniformly distributed in the open fire. The corners and edges of the tool are readily over heated, and subsequently in hardening break off. The tool, as a rule, is also not uniformly heated, one end or the centre showing a somewhat

FIG. 3.

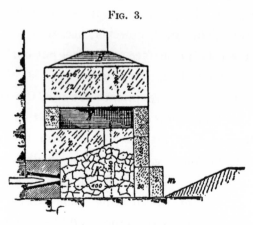

brighter heat than the other portions. In this case the open fire can, with small expense, be readily converted into a provisional furnace as shown in Figs. 3 and 4.

Set up bricks, in the manner shown in the illustration

FIG. 4.

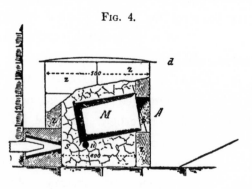

(Fig. 3), around the blast aperture so that an opening is left at *m* for loosening the fuel *K*. By laying a flat piece of iron across, an aperture is made at *A*, and after putting on the fuel, the structure is closed with a sheet-iron cover, which may be provided with a piece of stove-pipe for a chimney. By placing a few iron rods in the working aperture A, a grate may be made upon which the steel is heated without coming in contact with the fuel. During the process of heating, the aperture A is to be closed by a strip of sheet-iron placed upon the somewhat projecting grate-bars. In a provisional furnace thus arranged, thin, long tools are readily and well heated to a uniform temperature, especially with the use of charcoal as fuel.

For longer articles it will be necessary to arrange inside the heating space two tuyeres alongside each other and brick them over.

Unequal heating or over-heating of some teeth is an especially disagreeable feature in heating cutters, and cannot always be prevented even with the greatest care. However, large cutters are scarcely ever hardened from an open fire, and when it is done it is at the risk of the expensive tool.

Small cutters, punches, etc., as well as small tools in general, which are to be hardened all over, can be safely heated in a muffle built in the open fire as shown in Fig. 4.

Construct from old sheet-iron a pot of a size adapted to the articles to be heated and line it with clay mixed with cow-hair. Build up bricks around the tuyere and arrange the muffle as shown in Fig. 4. The back portion of the muffle is placed upon a piece of iron coated with clay which rests upon the bricks or upon a piece of brick cut so that the blast is divided or diverted into two lateral currents.

The mouth of the muffle is closed by a piece of sheet-iron b and the top of the furnace by a sheet-iron plate d.

It is of great advantage if the muffle M can be fixed between two tuyeres. The fuel is introduced from the top, the cover d being removed for that purpose. An aperture for loosening the fire is provided on the side. Of course the articles to be heated must not be brought into the muffle before the interior of the latter has acquired a suitable uniform temperature.

The devices for open fires shown in Figs. 3 and 4 are practicable with coke or charcoal, but not with coal. If coal has to be used, one of the reverberatory furnaces to be described later on may be employed for uniformly heating large pieces of steel or tools made therefrom.

Mild soft coke, such as is formed on the grate by the combustion of caking coal, gas-house coke in small pieces, etc., which are of a dark color, stain the hand, crumble readily, and when struck do not emit a clear sound, are not suitable for forging and hardening from open fires. Tools which come only partially in contact with the fire and the manufacture of which is simple and rapidly effected, such as hand and cross-cut chisels, pointed chisels, drills, hammers, etc., can, without hesitation, be forged and hardened from open fires with such coke. However, for all open fires, charcoal is without doubt the most suitable fuel. It readily yields to a greater extent the required degrees of temperature without a more abundant application of blast, and never contains foreign admixtures injurious to tool-steel. Hence charcoal should be exclusively used as fuel when tools of a finer quality are to be forged or hardened from an open fire, or when steel of special hardness, and hence more sensitive to fire, is to be subjected to these operations.

The great advantage of the use of charcoal for open fires is due to its purity and ready combustibility, and further to the fact that the degrees of temperature developed in the fire can be more readily observed from the exterior than with the use of any other fuel.

The great combustibility of charcoal allows of the use of the smallest quantities of blast, and hence greater protection against injurious effects to the steel is afforded. The action of the blast upon tool-steel or the finished tool produced from it, is the more injurious the higher the temperature at which it takes place, and the longer it lasts.

The disadvantages resulting from such action may be stated as follows :

1. The steel oxidizes on the surface and sinter is formed, and where oxidation has penetrated more deeply uneven spots result, so that the steel, especially in finished tools, presents a bad appearance, the smooth surface having been destroyed.

2. The oxygen of the blast readily withdraws from the surface of the hot steel, especially on the edges and corners, a portion of the carbon, the steel losing thereby considerably in hardness and cutting power.

3. When the temperature produced in the open fire is high and a great quantity of blast is present, the oxygen of the latter may not only combine with the iron and carbon on the surface of the steel, but may also penetrate more deeply into the latter, and by forming oxides with the iron, manganese and silicon of the steel cause a severance of the structure of the

latter.   Such steel is burnt and shows on the edges cracks of various depths.

In the subsequent forging, such steel crumbles on the places where the cracks are, and also in hardening the steel cracks.   There is no way of protecting the steel from the action of the blast in an open fire.   Such protection must be given by the mode of working, which should be as described below :

Heat the cold tool-steel or the finished tool by placing it upon the surface of the fuel, turning it frequently, and then push it into the portion of the fire showing the lowest degree of heat, hence near the circumference.   Allow the steel to remain here until it shows a uniform dark-red heat, and then push it into the hotter zone towards the centre of the fire.   To be sure, in this place the blast strikes the steel, but the latter having been preparatively heated, the action of the blast produces but little effect, except when the pieces are of such large dimensions as to prevent rapid heating.

If the effect of the blast is to be weakened, the current of air is reduced by placing in front of the nozzle irregularly-formed fragments of fire-brick, or the steel is protected by pieces of sheet iron arranged between the nozzle and the steel.   In the latter case it is necessary first to put on a large quantity of fuel and bring it into a red heat.   The attacks of the air upon the surface of the steel may be lessened by dipping the steel before heating in milk of lime or clay water (1 lb. of clay to 1 quart of water thoroughly stirred together).

Although open fires are much used and are very convenient for most purposes in the manufacture of tools, their disadvantages make themselves very much felt in the production and hardening of large complicated tools, and in the manufacture of implements in large quantities.

When a large number of tools, for instance, chisels, drills, picks, etc., are to be made or to be repaired, a dozen of them are at one time put in the fire with the intention of preparatively heating them, but the result is that they are not uniformly heated, they being partially overheated and even burnt. Such tools will in use prove not uniform and in consequence there will be loss of steel, as well as but slight efficiency.

Hence in factories where these causes have been recognized, it has, as a rule, been endeavored to arrange the heating devices so that the defects and imperfections due to the heating of the steel are in the main avoided. By such devices the strain on the tool-smith is relieved and limited to what is most essential, and fewer demands are made on his skill. Beside, as has been frequently mentioned, the efficiency of a tool carefully made is greater and consequently there is a saving in material and wages. The greater expense caused by the selection of better devices for forging and hardening are rapidly and fully made up by what is saved in wages.

Before entering into a discussion of the furnaces or fires to be used in separate cases, the great importance, especially in hardening, of the influence of light in judging the degree of temperature of the steel may here be referred to.

If a bar of steel, such as has been described on p. 21, be heated, the degrees of heat mentioned in the table facing p. 21 can be observed on it. These observations are based upon the purely subjective perception of the observer, ac-

3

cording to the relation between the heat and surrounding light. The heat of the same bar of steel will in absolute darkness present a different picture from that in dispersed day-light or in bright sunshine. What in the first case seemed to be bright red, appears cherry-red in the second case, and dark red in sunshine. By a changing light the eye is dazzled and becomes uncertain in judging the degrees of temperature.

By bearing in mind that the degrees of temperature to be used in forging, and especially in hardening, lie within very narrow limits and must be exclusively measured by the eye of the forge-man or hardener, it will be readily understood that not too much should be expected from these artisans if forced to work in a light room or in a much changing light. The advantage afforded to the workmen in this respect is to the interest of the consumer of the tools made by them.

Such advantage can readily be afforded by locating the devices for forging and hardening tools in the portion of the works which, for want of light, are not suitable for other purposes, or if this cannot be done, to prevent the entrance of a bright and changing light through the windows by providing them with a coat of paint, or curtains, etc.

As previously mentioned, heating the steel in an open fire has many drawbacks, which mainly result from the uneven temperature produced, the direct contact of the steel with the fuel, and from the blast. To avoid these drawbacks or to weaken their effect, furnaces are constructed in which the steel can come in contact only with the heated combustion-gases of the fuel, but not with the fuel itself or the blast.

As regards their arrangement, the different kinds of furnaces are constructed alike and are chiefly intended for fuels yielding no actual flame. The furnace previously described and illustrated (Fig. 2) is the most simple type.

The low degree of temperature which must be produced in such furnaces for heating tool-steel allows in most cases of their being kept in operation without the use of blast if a chimney of sufficient height—10 to 13 feet—is provided. One chimney may be used in common for several furnaces.

The dimensions of the furnaces described below are calculated for heating articles of medium size. They may be built larger or smaller according to the purpose they are to serve, and the dimensions must then be determined in proportion to the measurements given in the illustrations.*

Fig. 5 shows a furnace to be operated with coke, though a mixture of coke and smith-coal free from sulphur may also be used.

The furnace is built of fire brick and consists, commencing from below, of the ash-pit C, the grate r, the belly k, the feeding funnel M, and the working space A, the latter being all around provided with a ledge which serves for resting upon supports, articles which are to be heated without being held by tongs. The working space is terminated by a small vault provided with two to four symmetrically arranged apertures c d. These apertures are accessible by means of a door at M. They serve chiefly for the purpose of effecting a uniform distribution of temperature in the working space A, and for this purpose can be partially or

---

* The measurements given in the illustrations refer to centimeters—1 centimetre $= 0.394$ inch. A comparative table of centimetres and inches will be found on the end of the appendix.

entirely closed by the covers *a b.* The top of the furnace is
closed by the arch *C,* which carries the chimney *E.* If con-

FIG. 5.

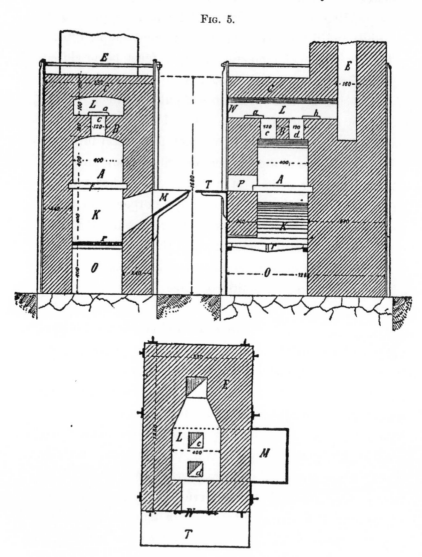

nection is to be made with another chimney or with one
used in common for several furnaces, the furnace-gases are

conducted into it through the space L, by a flue branching off sideways.

To facilitate handling the tool to be held in the furnace, a table T is placed in front of the working-hole P. Both P and the ash-pit C are to be provided with well-fitting sheet iron doors. The brick work of the furnace is held together by clamps and bolts as shown in the illustration. For regulating the draught the chimney-flue is provided with a special damper or the chimney itself with a movable cover. The heating of such furnaces requires from one to two hours according to their size, and is continued until the interior of the working space A, as well as its walls, shows a uniform bright red heat. The articles to be heated may now be introduced, the following general rules being observed. As the temperature rises considerably while the furnace is in operation, work should be commenced with smaller pieces which are most readily heated and more exposed to the danger of over-heating, and then pieces of larger dimensions may be gradually taken in hand. The heat which is given out by the glowing fuel being greater than that radiated from the furnace walls, the articles held in the furnace are heated more on one side and must therefore be frequently turned. When articles in a cold state are introduced, fresh fuel should at the same time be supplied, the higher heat of the furnace being thereby slightly modified, and with the use of smith-coal and coke mixed, the sooting flame developed affords some protection against over-heating and oxidation of the edges and corners of the steel. The tool-steel to be heated is either directly held with forge-tongs—the latter resting upon the working table T—or is laid upon a grate-like sup-

port. Small tools which are to be heated in quantities at a
a time may be brought into the furnace upon a sheet-iron
support by which they are protected from the direct heat of
the fuel.

FIG. 6.

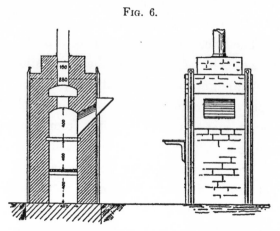

Fig. 6 shows a furnace of the same system as Fig. 5, but
which is operated with charcoal. The only difference is in
the arrangement of the door for the introduction of the fuel,

FIG. 7.

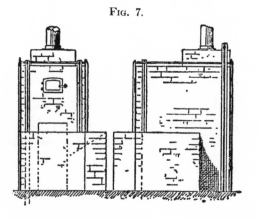

it being placed somewhat higher to avoid useless consumption of charcoal by combustion inside of the filling-box.

Fig. 7 shows the same furnace illustrated by Fig. 5, but to save iron parts the filling-box and working table are of brick work.

Fig. 8 shows a furnace for heating long articles, for instance long shear-knives, saws, etc., for forging and chiefly for hardening.

The regulating apertures, c d, and their covers, a b, with

Fig. 8.

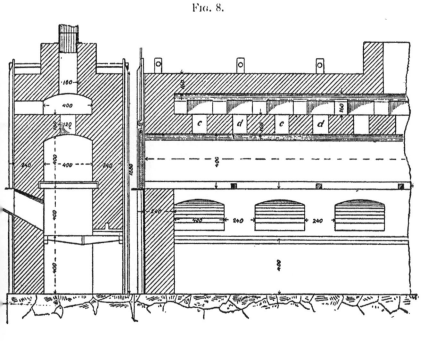

which all the furnaces are provided, as well as the chimney-damper, are of special importance for operating this furnace. By opening or closing the damper the draught in the furnace, and thereby the temperature in it, is increased or decreased, and by partly closing the apertures c d, the heat in separate parts of the furnace may be raised or lowered. Thus by closing the apertures c d in the left half of the furnace (Fig. 8), the heat is lowered in this portion and raised in the right half.

If the previously-described furnaces are to be fitted for coal, or if, with the use of hard coke, the articles to be heated are to be protected from a possible contact with the sharp-pointed flame developed thereby, the construction is as shown in Fig. 9.

Fig. 9.

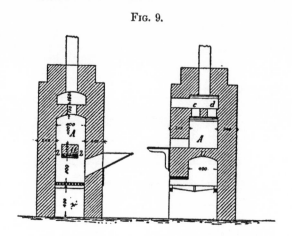

The coal or coke is here burnt under the arch U, the working space A being heated through the two slits Z. Thus the steel to be heated does not come in direct contact with the fire or the combustion-gases of the furnace, heating being effected under conditions similar to those in a muffle.

By constructing, about 10 to 16 inches above the arch U, another arch accessible by the door T, as shown in Fig. 10, the hearth thus obtained may be used for preparatory heating. When room is wanting, the use of such furnaces can be recommended especially for continuous work in forging and hardening articles on a large scale, for instance, in forging balls, files, shears, knives, etc., or in hardening calks for horse-shoes, parts of bicycles, etc.

By placing a muffle of iron or clay upon the arch U in Fig. 9, or two muffles, one above the other, as shown in Fig. 11, this furnace is converted into a regular muffle-furnace, the upper muffle serving for preparatory heating. In the muffle-furnace the steel is exclusively heated by the heat radiating from the sides of the muffle. However, it must not be supposed that tool-steel cannot be overheated

FIG. 10.   FIG. 11.

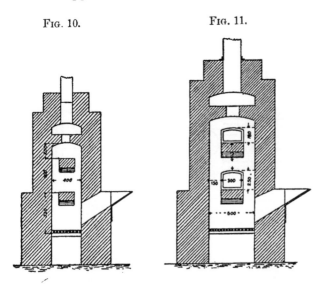

in the muffle; on the contrary, this is very easily done if due care is not observed. When a muffle furnace has been in operation for some time, so that the walls of the furnace and the muffle are finally heated throughout to a high degree, the heating of the steel is also effected at a higher temperature than permissible, and the steel is overheated, or at least the projecting corners and edges.

The interior of the muffle receives heat from every portion of the walls, but the greatest heat from the portions nearest to the fire-place, therefore from the bottom of the muffle, or from the bottom and a side wall. Now if the steel to be heated is brought directly in contact with the walls of the muffle, the portions touching the walls will acquire a higher temperature than the rest and may readily be overheated. Hence, the steel is placed upon supports— iron rods, or still better, cut pieces of fire-brick—so that it does not come in direct contact with the walls of the muffle and lies as nearly as possible in the centre. Uniform heat-

FIG. 12.

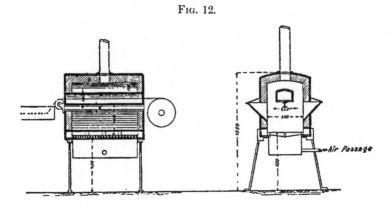

ing is promoted by frequently turning the steel. To prevent the access of air, the mouth of the muffle should be closed as tightly as possible by a door. Iron doors being readily warped by the heat, are not as good for this purpose as sliding doors of chamotte. If an iron door is used it should be provided in the centre with a looking-hole, which can be closed, for observing the temperature prevailing in the muffle; the looking-hole in the chamotte door is tightly closed with isinglass.

Fig. 12 shows a muffle-furnace to be worked with charcoal. It serves for continuous heating of band-steel for forging, hardening and eventually for tempering.

Muffle-furnaces, the muffles of which are heated by gas, are very cleanly, and easily attended. Fig. 13 represents such a furnace. The muffle, M, sits in an iron box, K, which is lined with chamotte. Underneath the muffle lies a heating pipe provided with a large number of holes arranged in several rows, and connected by means of a pipe with a small blower, V. The latter running at a high speed sucks gas from a gas-conduit with which it is con-

FIG. 13.

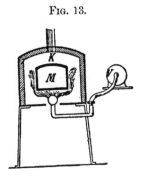

nected, mixes the gas with air and conducts the mixture to the heating pipe. The flame developed plays around the muffle and rapidly heats it to a uniform temperature, which by regulating the flow of gas can be readily raised or lowered.

Muffles of refractory clay must be heated very slowly and carefully, otherwise they may crack. Should cracks nevertheless be formed they may be repaired with a paste consisting of 4 parts graphite and 1 part clay.

The above-described furnaces are much used in practice, and when properly attended always prove satisfactory.

The cost of construction is not great, so that their use may be especially recommended in works for hardening expensive tools where the latter are still heated in the open forge-fire.

Although very seldom used for hardening, a reverberatory furnace is to be preferred to a larger number of forge-fires for continuous forging operations on tool steel.

Fig. 14 shows an ordinary forging reverberatory furnace

FIG. 14.

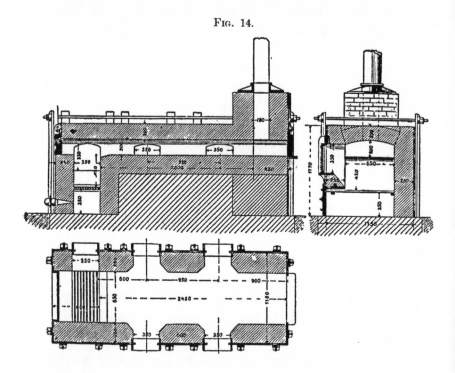

for heating with coal or lignite. It is provided on each side with two working doors and is, therefore, suitable for supplying several working places. Combustion is promoted by a blast-pipe terminating below the grate.

Fig. 15 shows the same furnace, the only difference being

that combustion of the coal is effected upon a step-grate with free draught.

FIG. 15.

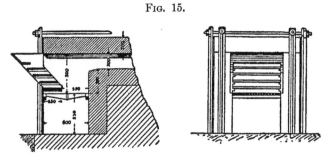

Fig. 16 shows a small reverbatory furnace such as is used for forging, and also for hardening operations on a small scale.

In case room is wanting, and also for small forging opera-

FIG. 16.

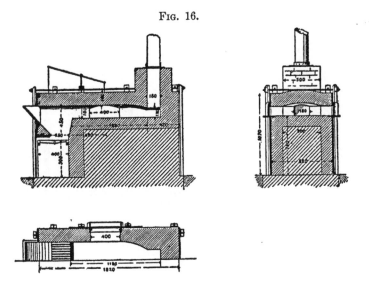

tions, the reverberatory furnace, Fig. 17, with two hearths may be used, the upper hearth serving for preparatory heating of the articles.

FIG. 17.

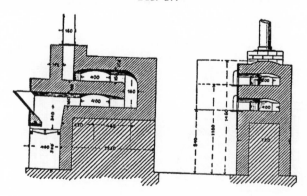

FIG. 18.

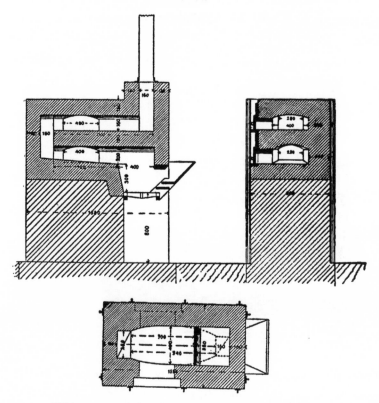

FORGING AND HARDENING REVERBERATORY FURNACE.

It must be borne in mind that the heat in a reverbatory furnace is not uniform, it being greater towards the fire-place and less intense near the working doors. Hence the steel is first brought into the colder portion of the furnace and then slowly pushed towards the hotter portions.

Articles to be heated for hardening are brought into the portion of the reverberatory furnace which shows the temperature to be used. To avoid contact with the highly heated bottom of the hearth, the articles are placed upon small iron

Fig. 19.

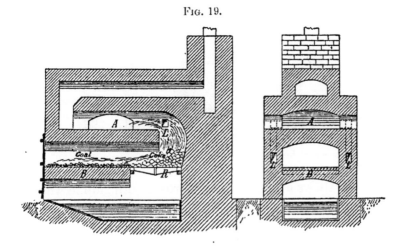

rods or pieces of fire brick, and they are protected from the direct action of the flame by placing a piece of sheet-iron bent at a right angle in front of them.

Fig. 18 shows a small reverberatory furnace with step-grate and two hearths, one above the other.

Fig. 19 illustrates a reverberatory furnace with a fire-place patented by Gasteiger, of Vienna. It has to some extent been introduced in practice on account of its great economy in the consumption of fuel. Its construction is according to the following principles:

The coal is piled upon the solid bottom, B, and gasified. The gases generated are highly heated by passing over the glowing coke upon the grate, R, and meeting the air, L, which enters from the sides, are burned in the space A.

The heat thus produced is intensified by water-gas, which is generated by steam evolved from a vessel filled with water which is fixed beneath the grate, the steam passing through the layer of glowing coke upon the grate R.

Fig. 20.

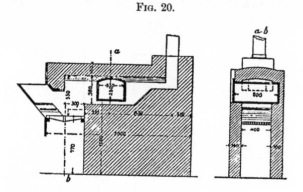

Fig. 20 shows a reverberatory furnace with muffle which is extensively used for hardening and annealing.'

VI.

APPLIANCES FOR ANNEALING STEEL.

As has been mentioned in a previous chapter, the structure of tool-steel is profoundly affected by mechanical working, the grain being finer the more vigorously the steel has been worked and the lower the temperature at which the work has been done.

Together with this alteration in the structure of the steel its properties of strength also undergo a change, the density being increased, the natural hardness enhanced and the capability of being worked decreased.

When in forging tool-steel, some portions of it are subjected to more vigorous mechanical working than others, and some portions remain almost untouched, the tool made from such steel showing in these places different degrees of strength as well as a different structure, and is said to have *forging strains*. In the further manipulation of such tool it works very unevenly and generally cracks in the subsequent hardening.

To take these strains out of the forged tools or to improve the capacity of crude steel for being worked, they are annealed before being worked.

In annealing tool-steel the same attention and care must be observed as in all other heating operations.

The following general rule applies to annealing :

The steel should be heated as slowly and uniformly as possible to a cherry-red heat and kept at this temperature until it may be supposed to have acquired a uniform heat throughout. It should then be protected against too rapid a loss of heat and allowed slowly to cool.

When, in annealing, tool-steel is exposed to the action of the air it becomes coated with a layer of oxide which frequently penetrates to some depth. The steel may even lose carbon on the surface, the decarbonized places acquiring no longer a sufficient degree of hardness, and at a high temperature the steel may readily be overheated and burnt.

Hence to protect tool-steel from the action of the air it is frequently annealed in vessels. Overheating in annealing

4

can only be prevented by scrupulous observation of the temperature.

Regarding the practical execution of annealing the following hints may be given :

Single tools or pieces of steel may be annealed in any of the furnaces previously described. When heating has progressed so far that the articles show a uniform red heat, they are imbedded in charcoal so as to be completely covered, and allowed slowly to cool. For annealing articles of steel of special hardness sheet-iron vessels, or cast-iron boxes especially made for this purpose are used. The articles are packed in the boxes between thoroughly burned charcoal, horn shavings, iron shavings free from rust, and the like. The covers of the boxes and any cracks are luted with clay, and the boxes are then heated, the same precautions as with steel not packed being observed. When sufficiently heated the boxes are covered with charcoal and allowed slowly to cool. The annealed articles should be taken from the protecting covering only when thoroughly cold.

For annealing larger quantities of tool-steel or manufactured articles, it is advisable, especially if the operation is to be frequently repeated, to construct special annealing furnaces.

Fig. 21 shows an annealing furnace for forged pieces, for instance cutters, files, swages, parts of bicycles, etc., or for tool-steel in short pieces. Wood and peat are chiefly used as fuel. Fuels yielding a quick heat should not be employed.

The pieces of steel or the articles are piled upon the hearth-bottom so that they are at a distance of a few centimetres from the openings *m m m;* in no case should they extend beyond them. The working door is then closed or

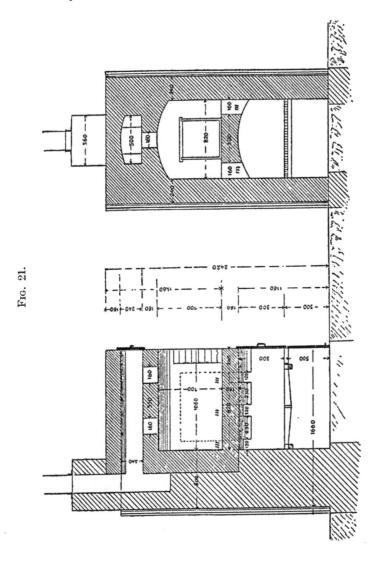

FIG. 21.

the door-opening bricked up with fire brick, the joints being thoroughly filled up with clay. For observing the temperature in the furnace looking-holes furnished with slides are arranged in the door and side walls.

The furnace is first heated with small quantities of fuel, firing being increased as the temperature rises until the steel shows throughout a uniform, not too bright, cherry-red heat. This temperature is attained in four to eight hours, according to the capacity of the furnace. All the flues serving for regulating the work of the furnace are now closed, as well as the chimney-damper and the doors, and cracks through which air might penetrate into the furnace are filled up with clay. The furnace is then allowed to cool slowly, which requires 48 to 72 hours.

Tool-steel thus annealed is of a very uniform, soft quality.

FIG. 22.

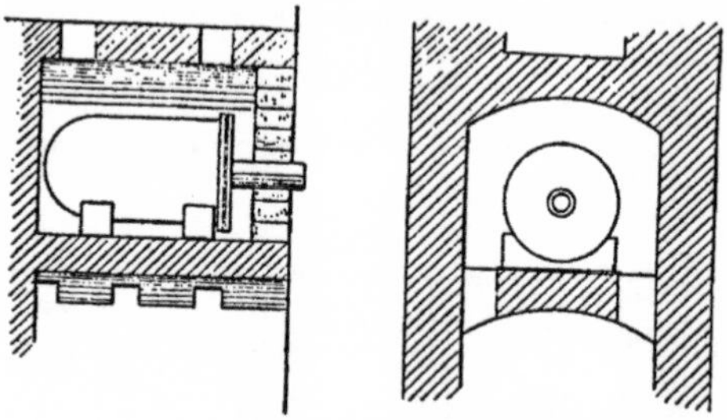

The same process of operating is to be observed in annealing steel in any of the furnaces previously described.

The furnace, Fig. 21, may also be used for annealing steel in an annealing pot or in annealing boxes, the latter when filled being placed upon the hearth-bottom and heated in exactly the same manner as above described. In this case coal may also be used as fuel.

The mode of arrangement will be seen from Figs. 21 and 22.

Fig. 23 shows a furnace with a large annealing pot, and Fig. 24, one with a larger number of smaller annealing pots.

FIG. 23.

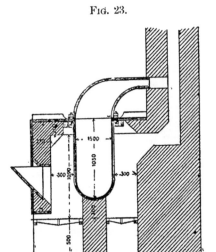

Regarding these furnaces, it may be mentioned that the annealing pots may readily be unevenly heated by the surrounding flame, in consequence of which the articles

FIG. 24.

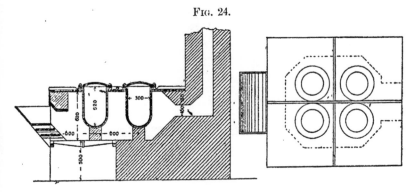

will be partially overheated, and in the subsequent hardening yield much waste. However, by very slow heating and constant attention this trouble can be avoided.

For the control of the result of the annealing operation, small rods of steel—waste or ends of the same quality of

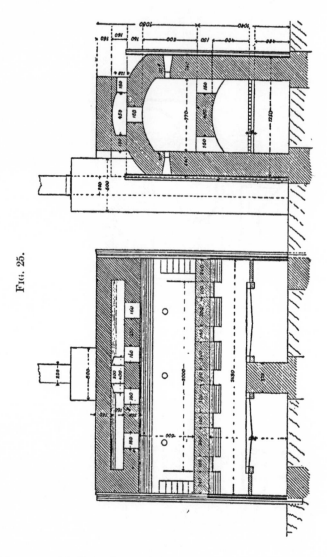

Fig. 25.

steel—are added to the articles to be annealed and placed in the parts of the furnace or annealing pots where overheating may readily take place. When annealing is finished, the rods are broken, and by comparing the resulting fracture with that of crude steel, a conclusion may be drawn as to the temperature attained in the annealing furnace.

Fig. 25 shows an annealing furnace for annealing long articles, of the same system as that represented by Fig. 20.

VII.

APPLIANCES FOR HARDENING STEEL.

THE various furnaces shown in Figs. 2 to 19 inclusive serve also for heating steel for hardening.

In no other operation in the manufacture of tools is thoroughly uniform heating of such importance as in hardening. The slightest error committed here causes damage which cannot be repaired.

Great care should therefore be exercised in the selection of furnaces for hardening, since in case the tool is spoiled, there is the additional loss of the cost of producing it.

Unfortunately there are no perfect appliances by means of which reliable heating of the tool can be automatically effected, the success of the operation depending in this respect always on the experience and attention of the hardener.

Imperfect heating appliances require closer attention and the observance of numerous rules, which soon tires the hardener, and many contingencies arise, the result being defective, or entirely useless, tools. Hence appliances should be provided which do not require such close attention as to make too great a demand on the skill of the hardener.

In reference to the choice of suitable appliances, the following hints may here be given :

If for hardening tools no other contrivance than an open fire is available, charcoal should be exclusively used for heating, even if it is thought that for momentary use other fuel might answer. The result of a change of fuel in favor of charcoal will in a short time show itself in the better quality, greater uniformity and longer endurance of the finished tool.

The devices shown in Figs. 3 and 4 may in many cases be employed with good results, for instance, for occasionally hardening fine tools, such as twist drills, broaches, cutters, etc.

By reason of their simple construction and their small cost, the furnaces, Figs. 5 to 7 inclusive, may be especially recommended for hardening all kinds of tools. With the use of charcoal as fuel, hardening is readily executed and without great danger even for the most complicated tools.

The furnace, Fig. 8, should not be wanting in a workshop where long articles, especially shear-knives, are hardened.

The various muffle furnaces previously described and illustrated are of all the furnaces best adapted for hardening, and offer the further advantage that fuel may be employed, the use of which in other furnaces might be injurious to the steel.

Although in the muffle furnace the tool does not come in direct contact with the fuel or its gases, the access of air to

it cannot be entirely avoided. Besides, by the tool coming
in direct contact with the walls of the muffle, it may be unevenly heated, and hence still safer methods are in many
cases desirable. The tools to be hardened are then heated
so that the air is absolutely excluded, either,

1. In a bath of molten metal, or,

2. In a bath of fused salts of a determined melting temperature.

Before entering into the further discussion of these
methods, it may be mentioned that in a melted mass tools
may also be heated too high as well as too low. The mere
heating in it offers no guarantee against the above-mentioned defects, if the melted mass itself has been heated too
high or not uniformly. It can just as readily be overheated as the wall of the muffle or a piece of steel, and its
being in a fluid state does not prevent unequal heating, just

FIG. 26.

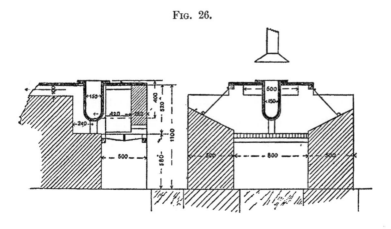

the same as with water which in a vessel may be brought
to boiling on the surface without raising the heat lower
down.

The metals or salts are melted in a cast-iron crucible, generally of a round cross-section, which is built in a furnace as shown in Fig. 26. The crucible is heated with coke yielding a mild heat, or still better, with charcoal.

Fig. 27 shows a furnace which may be operated with wood or coal, and also with lignite.

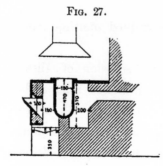

FIG. 27.

Without the use of a pyrometer, the temperature of the melted mass is controlled with difficulty, and can only be judged by the glow of the article heated in it. A pyrometer is also of great service for an annealing furnace.

Of the metals in a fluid state which serve for heating steel, refined lead without any additions is mostly used. Since lead melts at 637° F., and must be superheated to about 1382° F. in order to attain the required hardening temperature, it is exposed to strong oxidation on the surface, and the loss by evaporation is very large. To decrease this loss by oxidation, the surface of the melted metal is covered about ⅛ to ¾ inch deep with powdered charcoal. The lead vapors, which are very injurious to the respiratory organs, are carried off through a chimney indicated in Figs. 26 and 27.

Besides unequal heating or over-heating, the disadvantages which may arise from heating steel in melted lead are as follows: Impure lead containing sulphur yields a portion of the latter to the steel, the previously described soft spots being formed. Hence, to be sure, the freshly melted lead is for a few hours boiled before being used. When tools are heated in melted lead some of the latter readily adheres to some portions—in corners, on teeth, or other depressions—and prevents the hardening fluid from coming on these places in contact with the tool, and the latter here and there remains soft. To avoid this, cleanse the tools before hardening with benzine or alcohol from adhering grease, and brush them over with a mass of dough-like consistency prepared by mixing 1 part by volume of finely powdered charcoal (charred leather), 1 part by volume of rye flour and 1 part by volume of common salt with saturated solution of common salt in water.

Dry slowly and carefully the articles brushed over with this mixture before immersing them in the melted lead bath.

Although heating steel in fused salts may be designated as one of the best methods, it is not very widely practised, but where it is in use the great advantages offered by it are highly appreciated.

No absolute rule as to the mixture of salts to be fused can be given, it depending chiefly on the purpose it is to serve. Pure readily-fusible salts which, when heated above their melting points, rapidly volatilize are, of course, unsuitable, as well as salts which, by reason of their chemical composition, exert an injurious effect upon the steel to be heated in them.

On account of its special fitness and cheapness, common salt forms the chief constituent of the mixture. By the addition of a small quantity of readily-fusible soda (carbonate of soda), fusion of the common salt is accelerated, and the fluidity of the melted mass is increased by an addition of a small quantity of saltpetre. An addition of potassium chromate or of borax improves the properties of the bath, as well as an addition of yellow prussiate of potash, a possible injurious effect of the saltpetre, by decarbonizing the steel, being thereby prevented.

The salt is fused in a cast-iron crucible bricked in a furnace, such as shown in Figs. 25 and 26, the operation being as follows: Cover the bottom of the crucible, about one-third inch deep, with soda solidly rammed down, and fill the crucible up to the edge with common salt. Then heat the whole until fusion is complete. Now add gradually to the melt enough common salt sufficiently to fill the crucible, and then add about 5 per cent. by volume of saltpetre and 10 to 15 per cent. of potassium chromate.

Yellow prussiate of potash in small pieces is added to the melt as required, a larger quantity of it being used if the cementing effect is to be increased. In using yellow prussiate of potash it must be borne in mind that the vapors evolved are very poisonous and should be removed by means of a pipe.

As regards the condition of the articles to be heated, it may be mentioned that they must be free from liquid impurities; adhering oil or fluid fats must be previously removed by means of benzine or alcohol. Water should be particularly avoided, as the introduction of moist articles might readily cause an explosive ejection of the melted mass. Very cold articles should be warmed before being immersed in the melt.

The articles are suspended in the melt by small hooks, or they may be secured by ordinary iron wire. Articles which are to be partially hardened are held with tongs. which should be perfectly dry.

With this method it is of course also possible to overheat the melt, but this soon causes bubbling and ·running over, so that it can be more readily avoided than with the use of lead. However, unequal heating of the melt may readily occur. It may be detected by pushing a steel rod into the melt and observing the degrees of heating of various parts of it; the working of the furnace is regulated accordingly.

The disadvantages due to melted lead sticking to the tool are avoided with this method, the salt coating cracking off immediately from the tool when the latter is plunged in water.

The tools heated in melted salt retain their pure metallic surface after hardening. Articles which before hardening are coated with a layer of oxide, show after hardening a pure, smooth surface free from all scale. When to the salt melt larger quantities of yellow prussiate of potash are gradually added it exerts, in heating iron and steel, a strong cementing effect, so that wrought iron, otherwise not capable of being hardened, acquires, when treated in it, a surface as hard as glass, and the hardness of steel is not immaterially increased.

VIII.

HARDENING OF TOOL–STEEL IN GENERAL.

By hardening is understood the rapid cooling of steel from a distinctly perceptible red heat to a lower degree of temperature.

Before discussing the means which may be employed for cooling the steel to be hardened, it is necessary to refer to the changes the steel undergoes in hardening.

Steel when being heated for hardening must acquire a certain temperature if, after the subsequent rapid cooling, it shall possess hardness. The hardening temperature to be employed lies between 1292° and 1482° F., it being closer to the former for hard steel and closer to the latter for soft steel.

Hardened steel when gradually reheated to a higher temperature constantly loses in hardness, and has lost it entirely on acquiring a dark brown-red heat which is perceptible only in a dark room.

If the steel thus heated possesses a pure metallic surface, the latter, on heating, becomes coated with a film of iron oxide (ferroso-ferric oxide) which shows different colors at various degrees of heat. By this color—called *temper-color* —appearing upon the surface of the steel it may be judged what degree of temperature has been attained, and also whether heating has been uniform and sufficiently high for the desired degree of hardness.

With steel of different degrees of hardness, the temper-colors do not appear at exactly the same temperature. Hence their selection depends on the character of the steel used, but primarily on the degree of hardness previously given to the steel, and on the purpose for which the tool is to be used.

It is generally customary to temper the finished tools before they are used in order to lessen the brittleness in a hardened state and to impart to them the degree of toughness required to save the parts engaged in use from breaking.

For the reason previously mentioned, only general hints, which will be found in the table, facing p. 22, can be given for the selection of the temper colors.

For the appearance of the temper colors it is absolutely necessary for the pure metallic surface of the heated steel to come in contact with the air. Perceptible temper colors do not appear on impure surfaces, for instance, such as are covered with fat or other substances, nor on steel which, for the purpose of being tempered, is immersed in melted metal of a fixed temperature.

The changes which steel undergoes in hardening are chiefly due to the condition in which it was previous to this operation. By rapid cooling the hardening carbon which is formed by heating the steel becomes fixed and the structure of the steel concretes in the same condition in which it was at the time of hardening. This fixation extends also to the volume of the steel. Steel, like nearly all other metals, expands by heating, its volume being increased in the direction of definite dimensions. In hardening steel, *i. e.*, by rapid cooling in a lower temperature, heat is so rapidly withdrawn from it as to give the hardening carbon no time to be reconverted into carbide carbon, in which state the carbon occurs in unhardened steel.

The structure of the steel attains a state of rigidity and in this condition opposes considerably more resistance to a change in form than before, and in overcoming this resistance the component parts of the structure are not displaced but completely separated (torn, broken).

Hardened steel can no longer contract, and when once hardened also retains the larger volume which it possessed in the heated state. The fixed dimensions in the direction of thickness, and to a more limited extent, in that of breadth, are greater after hardening, while the length of hardened steel is less than previous to hardening.

These changes in the dimensions of the steel can practically be determined only with very accurate measuring instruments, though under certain conditions they become so great as to be perceptible to the naked eye. This may be observed on tool steel which has been repeatedly hardened without being annealed before each hardening.

When a piece of flat steel about 1½ inches wide, 0.39 inch thick and 3½ to 6 inches long, or a crude cutter about 1¾ inches in circumference and 0.47 to 0.59 inch thick, is hardened, then returned to the fire, and after carefully heating to the hardening temperature, again hardened, and these operations be repeated ten to twenty times, the changes in volume which the steel suffers are very conspicuous, and the changes will be the greater the harder the steel is. They are illustrated in cross-section in Fig. 28. For the experiment a steel plate with 1 per cent. carbon and 3.071 inches long, 1.732

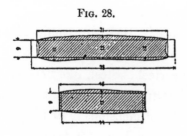

FIG. 28.

inches wide and 0.315 inch thick was used and hardened fifty-one times.

Repeated hardening could by no means cause such considerable changes in form if the steel every time it was heated previous to hardening would reassume its original dimensions. Such is the case only when the steel after heating is allowed slowly to cool. Simply heating the steel before hardening is not equivalent to annealing, if heating is not succeeded by slow cooling.

Strains which were in the steel before it is hardened are not removed by simply heating in hardening, but are even increased or enhanced after this operation. Special stress must be laid on this, because the incorrect view is very frequently held that heating before hardening is equivalent to annealing, and that such heating suffices for the compensation of the strains in the steel. As mentioned in a previous chapter, in working steel its structure undergoes a change, it becoming the denser the more vigorously the steel had previously been worked and the lower the temperature at which the operation had been performed. The steel decreases in volume while its specific gravity becomes higher. The component parts of the structure are then under the same influence as in hardened steel, i. e., in a state of rigidity which is much increased by hardening if the steel is not previously annealed. Hence there is greater danger of cracks being formed in consequence of the greater brittleness of steel not annealed.

The above described property of steel of assuming fixed dimensions in hardening operations immediately succeeding one another is practically utilized, for instance, with draw-plates. When the hole by use has been enlarged, the

draw-plate is hardened without previous annealing, the
result after each repetition of the process being a contraction of the hole.   If tool-steel of the same degree of hardness of, for instance, a round or square cross-section, be
notched at distances of about $\frac{1}{2}$, 1 and $1\frac{1}{2}$ inches from each
other, and then hardened and broken on the notched places,
the fractures allow of the following observations being
made :

The steel with the smallest cross-section will throughout
show a uniform, fine structure over the entire fracture if
the steel used for the experiment contained more than
about $\frac{3}{4}$ per cent. carbon, and heating before hardening
had been uniform throughout.   The heat from the interior
of the steel has been yielded up to the hardening fluid with
sufficient rapidity to produce uniform hardening throughout the entire cross-section.

On the fracture of the steel with medium cross-section
will be noticed a symmetrically arranged core of coarser
structure, which shows quite sharply defined boundaries.
The hardness of this core will be somewhat less than that
of the surface.   The delivery of heat from the interior
could not progress rapidly enough, and the hardness is not
so great as on the surface; hence the structure remains
coarser.

This phenomenon is still more strikingly perceptible on
the fracture of the steel with a larger cross-section.   The
interior of the steel shows, increasing towards the centre, a
coarser structure and considerably less hardness than on
the outer edges.

The above described phenomenon is seldom sufficiently
taken into consideration.   However, when considered in

connection with the fact that hardened steel increases in
volume, it is of the greatest importance for the practice of
hardening, especially as regards the mode of cooling the
steel to be hardened.

In Fig. 29 the dotted line represents the cross-section of
steel not hardened, and the full line the
cross-section of the same steel hardened.
By two concentric circles the surface is
divided into three parts, of which a represents the surface of greatest hardness, b that
of less, and c that of least, hardness.

Fig. 29.

Assume the zones thus formed to be continued throughout the entire piece of steel, spaces are formed inside
of which the steel is of approximately the same character. For instance, in the space a the steel acquired the
greatest hardness and in consequence also the greatest permanent expansion; in the space b the hardness as well as
the permanent expansion is less, the steel in consequence
of slower cooling having a tendency to contract somewhat
in this zone. In the space c the delivery of heat proceeded
still more slowly, and the steel has the least hardness and
the greatest tendency to occupy the original volume it
possessed before hardening, $i.\ e.$, to contract.

Now, if instead of two zones, we assume the interior of
the steel to be divided into innumerable such zones, it may
be imagined that each zone in its endeavor to contract,
exerts, commencing from the surface towards the centre, a
strain of constantly increasing force upon the outer portions
of the steel.

When the force of the strain developed in the separate
parts has become so great as to overcome the strength of

the steel, a displacement of the particles of its structure takes place, the steel suffering an expansion, whereby the force of the strain is checked or diminished. The greater the degree of hardness which the steel possesses, the less it will be capable of expanding, and the sooner a separation of the particles of the structure and consequently cracking will take place.

This cracking of the steel, however, does not always originate in the portions where the most powerful strain prevails, but in the portions of the steel which possess the least degree of extensibility and toughness. This is generally the case on the corners and edges, they being hardest, and seldom in the body of the tool, the softer core of which will expand without cracking.

The strains in the interior of the steel may, however, be so great that the limit of extension of which the less hard core is capable is exceeded.* The formation of the crack then commences in the interior of the steel.

While cracks commencing on or near the surface are, as a rule, formed early in the hardening fluid, and are perceptible immediately after hardening, severance in the interior generally takes place a considerable time after hardening and frequently causes the hardened tool to break only after several days.

Cracking which had its origin in the interior may frequently be observed on tools with a symmetrical cross-

---

* When the interior of the steel has flaws due to blisters or pipes, or when in consequence of liquidation, the steel is of varying chemical composition, the formation of cracks is the more sure to commence from the interior. The cause of this—a defect in the steel—is, however, plainly perceptible on the fracture.

section and on large dimensions, as well as when hard steel has been used, because with a symmetrical cross-section the centre of the steel is simultaneously subjected from several sides to a powerful strain, and in hard steel the strains are particularly great, its toughness being but slight.

Thin flat steel, profile steel of slight thickness, etc., are less liable to cracking in the interior than round or square steel, or cubes and balls.

In observing the fracture of hardened steel of a larger cross-section, it will be noticed that a narrower or wider border of an entirely uniform structure and equal hardness passes abruptly into an endless curved core of less hardness and of coarser structure. On the boundary between both

FIG. 30.

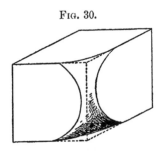

of them, close to the surface, are located the greatest strains, the steel possessing here the least toughness, and it cracks along the course of this boundary, hence generally in a curve. The corners of a cube, if the latter be repeatedly heated, will be severed, as shown in Fig. 30, in the first hardening if the steel is brittle or has become so by over-heating. The severance of the teeth of cutters also takes place in a line running in a curve along the boundary above described, and that of the face of a hammer as sketched in Fig. 31.

The force of the strain in forming cracks may frequently be so excessive, especially with hard steel, that when the latter cracks the fragments are hurled about with great force.

FIG. 31.

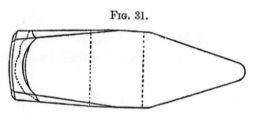

Fig. 32 represents a round bar of particularly hard special steel which, after hardening, has cracked from the interior. The force of strain exerted upon the interior of the steel may be judged from the manner in which the two halves are bent.

FIG. 32.   FIG. 33.

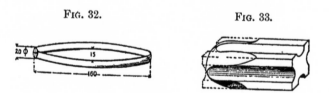

Fig. 33 shows a roll-cutter, the corners of which have been severed during the hardening in the direction of the above-mentioned curved line.

The above-mentioned forces are at work at every hardened tool and frequently cause cracking, even with the use of sound steel, if hardening has been done in a faulty manner, and it has been neglected to lessen the brittleness of the hardened steel by annealing at the proper time.

The previously described strains which in hardening are formed in the direction of the cross-section by an increase in the dimensions in the directions of the width and the thickness are highly influenced by the strains which are formed in consequence of the steel contracting in hardening.

The effect of these strains may be readily followed up by bearing in mind that the outermost layer of the steel which has been uniformly hardened has experienced a contraction, while the more slowly cooling core has a tendency to expand. In consequence of this, strains are formed in the layer of greatest hardness parallel to the direction of its length. These strains in the direction of the length cause a distortion of the steel if the latter has not been uniformly cooled in hardening. When a flat bar of steel is uniformly heated and the edge is lengthwise plunged in water so that about one-half of its width is cooled, it contracts on this side, while the uncooled portion projecting from the water expands. In consequence of these forces acting in a dissimilar manner, the steel acquires the form of a sickle, the

Fig. 34.

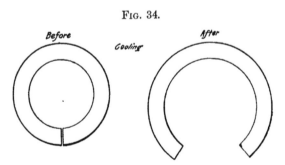

hardened portion curving inwardly and the unhardened portion outwardly.

If the bar of steel be bent together to a ring not entirely closed, and is then heated red hot, and the outer side of the ring is rapidly cooled, it will contract and cause the ring to open, see Fig. 34.

In hardening tools of an annular cross-section, the circumference becomes smaller, while the interior layers which have been less cooled endeavor to expand. The outer hard layer of the steel lies like a ring upon the interior layers, and the latter in their endeavor to expand make an effort to break the ring and frequently succeed in doing so. The cracks formed commence on the surface, and when once formed continue in the direction of the centre throughout the entire cross-section.

From this phenomenon we learn that by each heating, which in the interior of the steel acts upon layers already cooled, the endeavor to expand is increased, and the danger of cracking on the surface enhanced ; and the latter may even be first caused by it.

Cracking may be prevented by heating the steel from the outside, whereby the exterior hard layer is made tougher and the change in form can more readily take place.

The explanations given above of course refer only to tool-steel of normal composition, which can be well hardened. Steel less capable of being hardened when heated also suffers a change in volume, which, however, is not fixed by hardening, the hardness not penetrating to a sufficient depth, so that the hardened layer follows the re-expanding inner layers.

Manganese exerts a great influence upon the change in volume of the steel in hardening, and under certain conditions prevents it entirely. Ingot steel, still capable of being well hardened, with about 0.45 carbon and a higher content of manganese (0.8 to 1.0 per cent.) scarcely undergoes a change in its dimensions in hardening.

Hence such steel is used for tools on which no great demands are made and the dimensions of which must not be changed by hardening.

IX.

HARDENING OF TOOLS WHICH ARE TO BE HARDENED IN THEIR ENTIRETY.

FROM what has been said in the previous chapter, it is evident that the changes in volume caused by hardening and fixed by it are the immediate cause of cracking. By letting down the steel, i. e., by toughening it or making it more viscid, the particles of its structure can partially follow the changes in form and cracking is avoided. This toughening, however, must be done at the proper time, in fact, already during hardening, and hence belongs to the latter operation.

Below will be described the means by which failure in hardening may be avoided by letting down.

Before proceeding with the operation of hardening, the hardener should first of all form a clear idea of the portions of the tool in which formation of cracks is to be feared, and take such measures as may serve for the avoidance of them.

As a general rule it may be assumed that all projecting parts of a tool—corners, edges, teeth, etc.—will be most subject to snapping off or superficial separation.

Cracking from the interior is to be feared with massive tools of large dimensions. If such tools possess, in addition, projecting parts on the surface, as for instance, large screw taps, broaches, cutters, etc., the severance of these parts is the more to be feared the larger are the dimensions of the massive portions of the tools.

By letting down the steel on the surface where it possesses the greatest hardness, the particles of the structure acquire greater mobility; the condition of greatest rigidity being destroyed. In consequence of the greater toughness or viscidity thereby obtained, a scale-like separation of portions on the surface of the tool is avoided if letting down be effected at a time when separation has not already taken place.

As previously mentioned, cracking on or near the surface generally takes place already during cooling and is brought about the more readily the harder the steel is, the higher the temperature at which it has been hardened, and the colder the hardening fluid has been. If, for instance, a massive cutter with many teeth be hardened in water and allowed to cool only so far that when under water it can be touched with the hand, a crack will seldom be observed on the teeth. If, however, it be immediately returned to the cooling bath and allowed further to cool, the severance of single, and sometimes of all the teeth takes place very rapidly, and continues after the entirely cooled cutter has been taken from the hardening bath. When cracks on the surface are formed, a peculiar chinking noise well-known to experienced hardeners is heard. The snapping off of the teeth of the cutter is due to the fact that cooling off has been continued until the outermost, hardest layer has acquired the highest degree of brittleness and hardness, and the supply of heat from the interior was no longer sufficient to maintain it in a state of greatest toughness or viscidity.

The power of resisting the strain, which increases in force
with progressive cooling, becomes less and is finally overcome.

Superficial scale-like separations on tools are avoided by
interrupting cooling off at a period when the tool is completely hardened but not entirely cold (fractional hardening).

When the tool which is now allowed slowly to cool, still
retains sufficient heat in the interior to heat the surface to
a high degree, it may completely lose the hardness produced ; it becomes soft.

To prevent this, the partially cooled tool is not allowed
to get cold in the air, but is brought into a fluid which
withdraws the heat less rapidly than the hardening fluid—
generally oil or melted tallow—in which it is allowed
quietly to cool. However, this process cannot always be
relied upon, and in practice is reluctantly resorted to if
the highest degrees of hardness are positively to be obtained.

The highest degree of hardness, however, is only attainable when the tool is so completely cooled in hardening
that actual letting down from the interior is out of the
question.

By reference to the table facing p. 22 it will be seen
that the pale yellow temper-color commences to appear on
the heated steel at a temperature of 428° F. The tool
completely cooled off and hardened on the surface can
therefore without hesitation be heated from the outside to
about the above-mentioned temperature without fear of an
appreciable loss in hardness.

By this heating from the outside, the rigid hard surface

of the tool acquires a somewhat greater toughness or viscidity and can more readily follow the changes in volume of the interior without separating or losing in hardness.

This process is very frequently applied in practice by the hardener immediately returning the tool, when nearly completely cooled, to the hardening fire and at the same time heating it to a low degree of temperature at which an actual oxidation cannot take place. The reheated tool is then allowed slowly to cool. Great skill is, however, required for heating the steel very uniformly in the hardening fire, and preventing the sharp edges and corners from being finally overheated. Hence it is preferred to reheat the tool in hot sand or hot water instead of in the hardening or forge fire.

For this purpose the sand in a vessel is heated to such a degree that it can no longer be touched with the hand and small quantities of water poured upon it evaporate without hissing. The hardened tool is then at the proper time brought into the hot sand, and after being uniformly covered with it, allowed to cool. The manipulation with hot water is safer as regards the temperature to be used in reheating. The tool when almost completely cooled is quickly plunged in boiling or highly heated water and allowed to cool in it.

If the tool has been brought too soon into the hot water, *i. e.*, at a time when enough heat is still stored in the interior which in conjunction with the heat acting from the exterior would suffice to cause a decrease of hardness, the tool is after a few minutes returned to the hardening bath, again cooled off, and finally allowed to cool entirely in hot water.

The above-described process for protecting, during hardening, tools which are to be hardened in their entirety from possible cracking from the outside or inside by partial letting down or toughening, can be recommended for all kinds of tools without regard to their shape or size.

In some cases it is recommended to allow the hardened tool to remain in the hardening bath until it is entirely cold, and for a considerable time afterwards. The object of this process is to maintain undiminished the highest attainable degree of hardness in the interior of the tools. Letting down for the purpose of obtaining greater toughness or viscidity on the surface is then effected by the application of heat from the outside to the tool when completely cooled.

To be sure in using this process there is a risk of the loss of the tool. Scale-like separation of the corners and edges frequently takes place, particularly with the use of especially tough steel, though cracking of the piece from the interior occurs less frequently. The process finds practical application in hardening tools which work under especially high pressure. Large tools, for instance, rolls, or tools of especially complicated form, for instance, hollow bodies, tubes, etc., cannot be hardened by the process customary for smaller tools without fear of failure. For cooling such tools special devices are required, a detailed description of which will be given later on.

The hardening of tools the entire surface of which is to be uniformly hard, makes the greatest demands on the skill and experience of the hardener and requires the best kinds of appliances.

However in practice it does not alone suffice to protect one's self from the loss of a tool in hardening by heating and cooling in an appropriate manner in order to decrease the danger of cracking, but it is also necessary for the designer of the tool to take into consideration the influence of the forces which are formed in hardening and directed towards the destruction of the tool.

In some cases almost impossibilities are demanded from the hardener, the strangest shapes being, for instance, given to cutters without considering that the danger of the loss of the tool in hardening increases with size and complicated

FIG. 35.

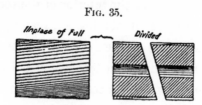

form, even if hardening has been done with the utmost care. Hence, tools of large or complicated shapes are, as far as their mode of application permits, divided into several sections, which when put together form the finished tool. Thus, for instance, long plane cutters are made in

FIG. 36.

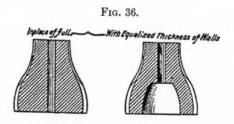

several short pieces, as shown in Fig. 35. The cutting surface is laid obliquely to the axis of the tool so that the joint is not detrimental to the cleanness of the work.

With thick irregular shapes, as shown in Fig. 36, the

danger of cracking from the interior in consequence of
unequal rapid cooling is diminished by boring out so as to
give the walls of the tool as nearly as possible the same
thickness. In other cases the strains in the interior of a
tool with symmetrical cross-section are diminished by boring it out. Very large taps, broaches, etc., which after
hardening readily crack from the interior and cool with
difficulty in consequence of the heat stored in the interior,

FIG. 37.

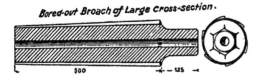

Bored-out Broach of Large Cross-section.

are bored through in the direction of their length, as shown
in Fig. 37.

In solid tools the strains formed in hardening converge
in the centre, while in tools bored out they are distributed
upon the circumference of the bore.

An enumeration of all the cases in which the danger of
failure in hardening may be diminished by a very simple
change in the form of the tool is, of course, out of the question, but it may be mentioned that in practice such cases
are quite numerous.

The main points to be considered in the construction of
a tool are whether its form is as simple as possible and
whether the projecting portions correspond favorably to the
bulk of the tool; further, whether sharp corners and edges
are as much as possible avoided, and whether there is a
gradual transition of different cross-sections.

X.

HARDENING OF TOOLS WHICH ARE ONLY TO BE PARTIALLY HARDENED.

THE operation of hardening tools which must possess throughout a uniformly hard surface, as described in the previous chapter, presents the greatest difficulties to the manufacturer of tools, and requires considerable skill and experience.

However, the greater number of tools are only partially hardened, namely, the parts which in use are subject to great wear, while the other portions are to be as tough as possible.

With reference to the main points and the nature of the tools, partial hardening may be divided into three groups as follows:

1. In heating the tool for hardening, every portion of it which is to be hardened may be heated to such an extent as is required to obtain a uniform degree of hardness, and the operation of hardening may also be carried on accordingly, for instance, with turning knives, chisels, gouges, etc.

2. On account of its small size or shape, the entire bulk of the tool to be partially hardened has to be heated, but hardening may be partially effected as, for instance, with hammers, sledges, small punches, short and broad-cutting knives, etc.

3. The tool must be heated and hardened all over to prevent distortion, partial cracking, etc. The hardness of the portions of the tool not subject to wear is

after hardening reduced by reheating, for instance, with all kinds of knives (shear-knives, wood-planing knives, stamping knives for paper and leather).

What has been said in reference to heating tools which are to be hardened all over, applies also to partial hardening, *i. e.*, the steel must under no conditions be heated too highly, or unequally on the places to be hardened.

Heating must also be so effected that the portions to be hardened show a uniform hardening temperature which in not too rapid gradation decreases in the portions of the tool not to be hardened.

When a tool runs to a thin edge the latter becomes warm in the fire and is readily overheated before the portion further back acquires the required temperature. In this case the portion back of the edge is first heated to a dark cherry red, and then the edge to the hardening temperature.

If the portion of the tool to be hardened possesses a small cross-section running rapidly and quite immediately into a larger cross-section, for instance, a punch, Fig 38, the thicker portion *a* is first heated until it

Fig. 38.

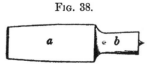

shows the proper hardening temperature, and then the thinner portion *b*.

Broad cutters may in heating be readily overheated on the sharp corners if not protected therefrom by occasional cooling off. Such cooling off, in case the corners should show a higher temperature than the other portions of the cutter, is effected by dabbing them with a moist rag or scattering upon them a dry powder consisting of:

6

Calcined common salt...................... 1 part by volume.
Pulverized hoofs .... 1 part by volume.
Powdered charred leather .... 1 part by volume.
Rye flour................................. 1 part by volume.

If, on account of their form, tools which are to be partially hardened have to be heated all over, it must be done so that the portion to be hardened acquires last of all the proper hardening temperature, the portion not to be hardened being, therefore, first exposed to the higher temperature in the fire. As regards tools which are to be partially hardened, it may be mentioned, that by cooling them a sharp line of demarkation between the hardened and unhardened portions should never be allowed to form, but there should be a very gradual transition of one into the other.

If a bar of steel be heated to a uniform cherry-red, and then immersed to a certain depth in water and cooled so that a sharp line of demarkation is formed between the hardened and unhardened portions, and an experiment be made in breaking the bar, fracture will take place the more assuredly on the line of the hardened part, the harder the steel used.

The strains on the sharp boundary of the hardened portions are sufficiently great to make such demands on the strength that less force is required to produce fracture along this line than on any other places of the steel. In use, tools thus hardened, soon break along the boundary between the hardened and unhardened portions, if it does not already occur in hardening.

A gradual progress of hardening is attained by slowly moving the tool up and down during cooling.

If tools are to be hardened so that there is a gradual transition of the hardened into the unhardened portions with rapid cooling, as is frequently desirable in the manufacture on a large scale, the appliances for heating must of course be made with great care, so that abrupt transitions of temperature do not occur on any portion of the tool, because when such is the case, partial hardening also causes a sharp line of demarkation between the hardened and unhardened portions.

XI.

COOLING OF TOOLS IN HARDENING AND DEVICES FOR THIS PURPOSE.

THE general rule for cooling tools is of exactly the same purport as that for heating them for hardening, namely,

Cooling of the red-hot tool should be effected so evenly that the heat is uniformly withdrawn from the portion to be hardened.

This rule seems very simple, but in practice it is not always easy to follow it, because even by simply immersing the red-hot tool in the hardening fluid, unequal cooling takes place at the outset.

While the tool remains in the hardening bath, steam is formed by the evaporation of the fluid, and where the steam cannot escape with sufficient rapidity the tool is enveloped by a layer of it which prevents uniform hardening. Hence, the article to be cooled must be kept in constant motion so that it always comes in contact with fresh layers of the hardening fluid, and the steam evolved can escape with greater facility. However, by this motion the separate sides of the tool are alternately brought into more energetic contact with the cooling fluid and also unequally cooled, the consequence being that the tool warps or cracks. When the tools to be hardened are small, heat is very rapidly withdrawn from them and unequal cooling by immersing and moving them about seldom causes warping or cracking.

When especially large and heavy tools are to be hardened, or tools which on account of their cross-section show a ready tendency to warp, they are not moved about in the hardening fluid, but the latter is set in motion. This motion may be a simple one, when the hardening fluid acts from one side upon the tool, or a combined one, when it acts simultaneously from several sides.

Before entering upon a further explanation of what has to be observed in the actual hardening of tools, attention may be drawn to a very important fact which may often be noticed.

It frequently happens that tools which have been heated with the greatest care are previously to hardening, exposed to cooling, and then hardened when in a state of unequal temperature. As a rule this is not taken into consideration and ultimate failure in hardening is ascribed to entirely different causes.

Unequal cooling of the tool immediately before hardening may be caused by a sharp draught of air, and this should be avoided as much as possible in the neighborhood of the hardening fire. When the red-hot tool for the purpose of hardening is brought from the furnace or fire, it is frequently laid upon cold iron plates or even upon a damp support to facilitate catching it with the tongs. The un-equal cooling thus induced is frequently the cause of cracks being formed.

Catching the red-hot tool with cold tongs, or with tongs wet from a previous hardening operation, will almost always cause the formation of cracks, especially with a tool of a small cross-section, since uneven cooling penetrates immediately, the entire cross-section. Such a tool is best heated by catching it with the tongs and heating it to-gether with the latter. When this cannot be done, the jaws of the tongs should be heated to a dark red heat each time before seizing the red-hot tool. Cracks which are .

FIG. 39.

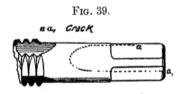

formed by uneven cooling from catching with the tongs, as a rule, run a quite regular course, they being formed either in the direction of the length of the tool, or return in a curved line to the initial point. The dotted line in Fig. 39 shows the course of such a crack on a twist-drill.

When a tool which is to be hardened is grasped with the tongs and together with the latter immersed in the harden-ing fluid, care should be taken that the jaws of the tongs come in contact with as few points of the tool as possible, because the portions of the latter covered by the tongs do not cool with sufficient rapidity; the consequence being uneven hardness, and cracks commence to form on the places grasped by the tongs. For catching tools to be hardened, tongs with jaws terminating in as sharp points or edges as possible should be used. Tools bored through are taken up with a hook pushed through the bore.

Fig. 40 shows a cylindrical revolving knife held by tongs properly selected; Fig. 41, a cutter, and Fig. 42, a cutter grasped with an iron hook. On the other hand, Fig. 43 shows a cutter held by tongs not suitable for hardening purposes.

In many cases when thin flat tools are to be partially hardened, the portions not to be hardened are covered before cooling with pieces of sheet-iron, etc., or the tools are grasped with tongs, the jaws of which are so shaped as to

| FIG. 40. | FIG. 41. | FIG. 42. | FIG. 43. |

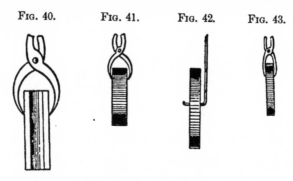

form the covering. However, cracking and distortion of the tool are frequently caused by this mode of partially hardening, and it should only be used with very soft varieties of steel which require merely superficial hardness. When thin flat articles, some portions of which have been covered, are heated and hardened, a sharply defined boundary, in accordance with the extent of the covering, is formed between the covered portions, which are not at all or only slightly hardened and the portions which are completely hardened. The thicker the covering is, the more sharply defined this boundary will be. In order to broaden the transition from the unhardened to the hardened portions, the covering should be thinner towards its edge, and the latter should not very firmly rest upon the article. However, when the same object can be attained by another method, for instance by reheating the hardened portions which are to be soft, it is to be preferred.

For thick tools of symmetrical form, unequal hardening by covering separate portions is employed when the strains formed in hardening are to be dispersed and not allowed to converge towards a centre; further, when the zone of greatest hardness is to be limited to as small a space as possible.

Fig. 44 represents one-half of a cutter which has been

FIG. 44.

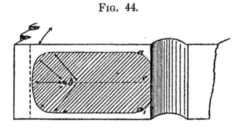

hardened without covering. The exterior zone which has acquired the full degree of hardness is not hatched, and is bounded by the line $a\,a'$. Along this line the steel possesses least strength and, in hardening, separation of teeth will take place in accordance with it. From the point b and the line c, the strains act radially towards the exterior. When in hardening the same cutter is protected on the face by covering with sheet-iron discs as shown in Fig. 45, cooling and hardening proceed as shown in the sketch of the cross-section. The line $a\,a'$ shows the course of the boundary of least strength. The strains disperse themselves in parallel direction and separation of teeth in the direction of their course does not so readily take place.*

Fig. 45.

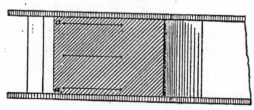

It is an absolutely necessary provision for this mode of protection that the covering should project over the portions to be protected. If the covering were of a smaller extent, for instance as shown in Fig. 46, the result would

Fig. 46.

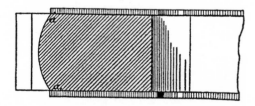

be just the reverse from that intended, because a sharply defined boundary would be formed between the hardened and unhardened portions of the tool.

> *The application of the above-mentioned method is of course very limited, and depends on the form of the tool as well as on the extent of the surface to be hardened.*

It is largely employed in hardening rolls of very hard

---

* Beside the above-mentioned strains in the direction of the cross-section, the entire circumference stands under considerable strain, which is caused by its having been shortened by hardening, and by the pressure of the inner less hardened layers upon the outer hard layer.

steel in order to decrease the danger of cracking, and in hardening parts of machines certain portions of the surfaces of which are to be hard to prevent rapid wear.

Tools which by reason of their forms or dimensions have for the purpose of hardening to be heated in their entirety, should always be cooled in a fluid which is in a state of motion in such a manner that the surface to be hardened is exposed to a continuous stream.

The most simple manner of effecting this is to immerse the red-hot tool in running water. However, uniform action by this means is only effected laterally and hardening in running water requires, as a rule, the immersion of the entire red-hot tool, consequently also cooling—hardening— of the portions not intended to be hard. Hardening in water running in a horizontal direction, for instance, in gutters, channels, etc., offers no appreciable advantages as regards uniformity of cooling, especially of larger surfaces, the only advantage being the relatively uniform temperature of the water on account of the latter being constantly renewed. The temperature of free running water is of course not absolutely uniform, it depending on the season of the year.

The mode of using running water for hardening is as follows: The water is conducted into the vessel to be used for hardening purposes, and the supply regulated so that the temperature remains the same even when hardening is continuously carried on. Fig. 47 shows a simple device for the purpose. The water runs in at Z and runs off through a notch A in the side of a barrel which serves as the hardening vessel.

For hardening tools, special advantages are offered by

the employment of falling water, because it is feasible thereby to expose to it only the portions of the red-hot tool which are to be hardened. The process of hardening by means of free-falling water is very simple, it being only necessary to expose the area to be hardened to the action of the jet of water until completely cooled. The main point

FIG. 47.

in this process is that every portion of the area to he hardened is uniformly struck by the jet of water. If the latter is not sufficient to cover uniformly the area to be hardened, it may be adjusted by moving the tool about, but the danger of defective hardening is not entirely averted.

With a plentiful supply of water the device shown in Fig. 48 is very suitable for hardening under falling water. The water is conducted to the vat H, which may be constructed of wood, iron, or brick, through the conduit L, the notch E serving for the overflow of the excess of hardening water as well as for keeping the latter at the same level. To the conduit L is fitted a cock the discharge opening of which is provided with a thread to which may be screwed a rose (see sketch Fig. 48). This rose comes into use when the water flowing from the cock M is to be distributed over a larger area. Of course with the use of the rose the water must flow under a certain pressure. The tool to be hard-

ened is laid upon the grate R fixed under the cock M, and
the latter being opened, cooling is continued until the tool
is cold.   In many cases, especially when very hard steel is
hardened under the jet of water, the strains formed by
hardening have to be somewhat equalized.   For this purpose the tool is not allowed to become entirely cold under
the jet of water, but hardening is interrupted at a suitable

FIG. 48.

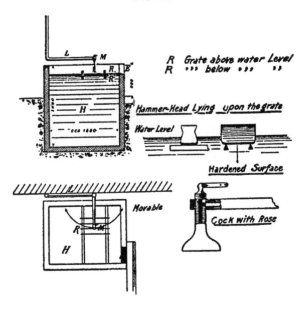

moment and the tool, hardened side down, is laid upon a
grate extending somewhat below the surface of the water,
upon which it is allowed completely to cool.   Oxidation of
the hard surface is prevented by fresh water running constantly in at the side (see sketch, Fig. 48).   The cock M
should extend to the centre of the hardening vat and should
be movable so that it can be turned out of the way when
not in use.

In case a water conduit or running water is not available, the device shown in Fig. 49 may serve the purpose.

It is, of course, only intended for occasional hardening under the water-jet, or when for want of water flowing under pressure, hardening is to be effected in an ascending water-jet.

The very simple device consists of two vessels *H H*, arranged one above the other. From the upper vessel *H*, a pipe *R* conducts the water to the cock *M*. The pipe *R*

Fig. 49.

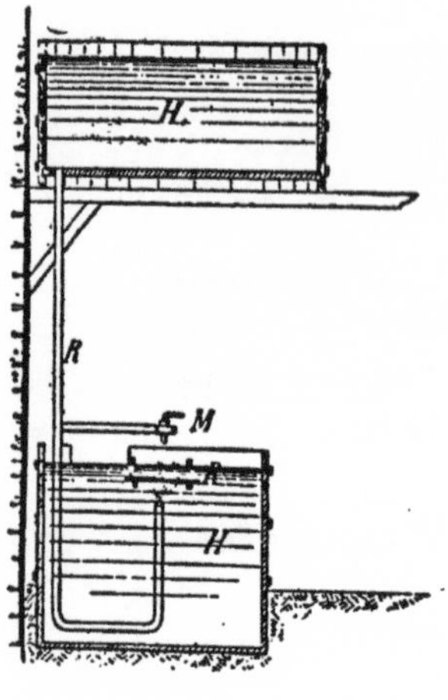

Fig. 50.

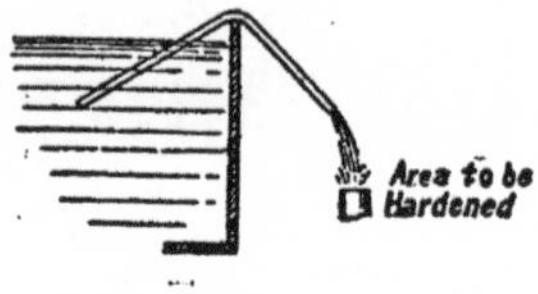

shown in the same sketch extends to the bottom of the vessel *H* and is then bent upward, thus causing the jet of water to ascend. When hardening is to be effected under an ascending water-jet and there is at hand neither a water conduit nor other device for producing a water-jet, a funnel constantly kept full by pouring in it a uniform supply of water may be used. Still more convenient is the use of a siphon made by bending a piece of gas pipe. For use the siphon is filled with water and brought into the position shown in Fig. 50. As the water flows from the siphon with but slight force, this device, of course, is available only for small areas.

The object of hardening in an ascending water-jet is the same as in a falling water-jet, and its use is preferred if permitted by the form of the tool or the area to be hardened.

A falling and an ascending water-jet are simultaneously employed in hardening tools which are to possess the same degree of hardness on two opposite sides, and one end of which, by reason of their short length, cannot be heated without the heat being at the same time transmitted to the other end, for instance, pivots, short hammers, sledges, etc.

While tool-areas may under all conditions be hardened by means of a falling water-jet—and in some cases must even be thus hardened—the employment of an ascending water-jet depends on the form of the area to be hardened.

When the area to be hardened is level or has elevations or bulges out otherwise, hardening may be effected with an ascending, as well as a falling, water-jet, but the former method is to be preferred, since greater uniformity of hardness can be attained with it. If, however, the area to be hardened has depressions, or curves inwardly, hardening with a falling water-jet is the only method available. The vapors formed in cooling with an ascending water-jet cannot immediately escape, and being constantly reformed, prevent contact of the water with the steel so that hardening is not effected and the depressions to be hardened remain soft. The device for hardening in an ascending water-jet is shown in Fig. 51. The supply pipe, R, is placed in the hardening vessel, H. The supply pipe terminates below the water-level, and the water flowing from it carries along quantities of surrounding water and conveys them to the area to be hardened. For the support of the tool, *W*, to be hardened, a grate, *R*, extending below the surface of the water, is fitted to the hardening vessel (see Fig. 51). When it is not feasible to support the tool upon the grate, for fear that the area to be hardened may

FIG. 51. FIG. 52.

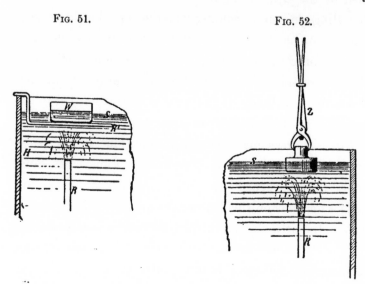

remain soft on the places which come in contact with the grate, it must be held with the tongs *Z*, and suspended free, as shown in Fig. 52. The tool to be hardened remains exposed to the ascending water-jet until it is cold, it being, of course, immersed to a sufficient depth. The more vigorous the flow of water is, the better the hardening will turn out.

For tools, the entire bulk of which is to be hardened, but which are of such large dimensions as to make even cooling by moving them in the hardening water no longer possible, special devices are required for the uniform conveyance of the hardening fluid from every side to the tool. A brief description of such devices for hardening balls and rolls will be found in the Appendix.

The method of hardening by means of a jet may also be employed, if the operation of hardening, as is customary in some factories, is not effected in pure water, but in a large reservoir which contains a sufficient quantity of water compounded with salts or acids to allow of a large number of tools to be hardened without its temperature being essentially raised. To be sure special small pumps are then required, which for occasional use may be worked by hand while for continuous working power is required. The advantage of the use of such pumps is that the lower cooler layers of the hardening bath are brought to the surface, the temperature of the bath being thus equalized.

Hollow bodies which are to be hardened inside, or inside and outside, require special devices so that cooling may take place without uneven hardening being caused by the development of steam.

Tools, entirely bored through or hollow, are cooled by conducting a powerful jet of water through the interior. If the surface is also to be hardened it must be cooled by a simultaneously acting water jet.

Fig. 53 shows the operation of hardening from the inside, and Fig. 54 from the outside and inside. These devices require small power plants.

The hollow body H is placed between two pipes, a b, Fig. 53. The pipe b is fitted movable to the pipe c so that the hollow body H can be firmly wedged between the two pipes. A supply cock being at the same time opened, water under high pressure enters and in passing through the hollow body hardens its walls.

Fig. 54 shows the hollow body also surrounded by a movable pipe d. By closing the pipes b, c and d, the water flows through between them and hardens the surface and interior of the hollow body. In place of the additional pipe d, the surface may also be cooled by direct-acting roses.

Cooling the interior of a hollow body which is not en-

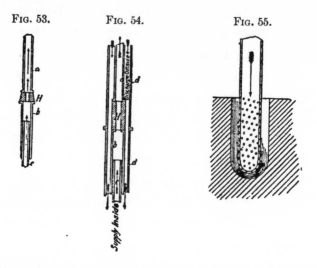

Fig. 53.	Fig. 54.	Fig. 55.

tirely perforated, presents greater difficulties, and the more so the narrower and deeper the opening is. Hardening is then effected by the direct introduction of as powerful a jet of water as possible. If the size of the aperture allows of the introduction of a pipe, one end of which, corresponding to the depth of the aperture, is provided with numerous small holes, even hardening is readily obtained with a powerful pressure of water (see Fig. 55).

XII.

LIQUIDS USED IN QUENCHING STEEL.

1. PURE WATER.

FOR quenching steel for the purpose of hardening, pure water is mostly employed, and the colder it is the more intense the hardening will be, and the less so the warmer it is.

The most suitable temperature of the hardening water lies between 60° and 72° F. Colder water will not appreciably increase the hardening effect, but decreases the toughness of the steel, the latter becoming very brittle.

It may be laid down as a rule that the larger the cross-section of the tool to be hardened, the lower the temperature of the water must be. For large tools a temperature of over 65° F. should not be used. Small, especially thin, tools will take sufficient hardness at a temperature of the bath of about 86° to 95° F. The temperature of the hardening water must also be chosen in accordance with the form and size of the tool, as well as the degree of hardness desired.

Chemically pure water, *i. e.* water which does not contain other substances, does not exist in nature, and its chemical composition varies very much, it containing soluble and insoluble mineral salts and acids in a state of minute division.

Many salts and acids increase the heat-conducting power of water and hence bring about more intense hardening, while earths and other mineral substances in the water have a detrimental effect in this respect. It is for this reason that

7

pure well or spring water which chiefly contains acids and carbonates in solution will harden more sharply than river water (especially when turbid) or water containing lime.

In order to lessen the too intense hardening effect of well or spring water, some soda or potash may be dissolved in it; and to impart better hardening qualities to river water, it is mixed with small quantities of acids (hydrochloric acid, sulphuric acid, vinegar, etc.). Experienced hardeners place great value upon the favorable effect of hardening water which has been used for some time, and are very unwilling to renew it, because it is claimed to improve in quality, *i. e.*, it hardens well and mild. The reason for this has to be found in the fact that fine particles of undissolved admixtures gradually separate on their own account and sink to the bottom, while by the action of the hot steel which is plunged into the water, soluble substances are gradually removed by evaporation as well as by conversion into insoluble substances, the latter being partially precipitated upon the surface of the steel. Hardening water which has been used for some time will finally get into such a condi tion that the unintentional admixtures have become harmless, and hardening in it always yields uniform results.

If this condition is to be brought about from the start, the water should be boiled before its being used for hardening, or it may, so to say, be sterilized by quenching in it larger quantities of hot pieces of iron.

2. Hardening Water Mixed with Soluble Constituents.

Soluble admixtures, as previously mentioned, exert great influence upon the action of the hardening water, their effect being either an intensifying one, if the heat-conducting power of the water is increased, or a moderating and retarding one, if the heat-conducting power of the water is decreased, or the boiling point is essentially lowered.

Common salt is mostly used as a soluble admixture for increasing the heat-conducting power. It is dissolved in varying proportions by weight, though generally a saturated solution of it is used. The latter may be recommended whenever tools of complicated shape which are to possess great hardness are to be hardened in large numbers or in rapid succession.

In using such cooling fluid, it must be borne in mind that the bath must be of sufficient volume so that its temperature is not essentially raised by continuous hardening. A vessel of as large a size as possible should therefore be selected for the reception of the bath, and it is better to use a shallow vessel of large diameter than a narrow and deep one.

Soda (carbonate of soda) and sal ammoniac dissolved in water do not produce such an intense effect as common salt; but though they are more seldom used they are excellent admixtures to hardening water, particularly for complicated tools, especially for cutters, a superficial separation of some parts of which may be feared.

Acids in particular intensify the action of the hardening water in a much higher degree than common salt. They may be added in quantities of up to 2 per cent., sometimes in combination with salts. Organic acids (acetic, citric acids) produce a milder effect than mineral acids (hydrochloric, nitric and sulphuric acids).

Acidulated water is used for tools which are to possess

the highest attainable degree of hardness (cutters for working articles of special hardness), or for giving a surface of sufficient hardness to steel which does not possess good hardening qualities.

Alcohol lowers the boiling point of water and, when the latter comes in contact with the hot tool, causes such an energetic evaporation that hardening is more or less retarded according to the strength of the mixture. Water which contains a large proportion of alcohol does not harden at all.

Soap. Soap water does not harden steel, and this property is made use of for rapidly quenching steel when the latter is not to be sufficiently hardened by cooling. When some parts of a tool which has been entirely hardened are to be made soft, they are brought to a red heat and cooled in soap water (tangs of files, knives, sabres, saws, etc.).

Soluble admixtures of organic nature retard hardening more or less according to the proportion of admixture, and hence modify the action of pure water. In practice they are rarely used and then only in small quantities, for instance, milk, sour beer, etc.

3. Hardening Water Mixed with Insoluble Constituents.

Such water is frequently used for hardening especially complicated tools to protect them against cracks. For its preparation, lime in the form of milk of lime is chiefly employed, clay or loam being more seldom used. Such water has a more or less retarding effect according to the strength of the mixture. The admixed constituent being intimately incorporated with the water, is precipitated upon the hot

steel when the latter is plunged into the bath, and forms upon
it a thin layer which prevents direct contact of the steel with
the hardening water. Cooling thus takes place more slowly,
and by reason of the retarding effect, the hardening is less
intense in character.

4. Hardening Water Mixed with Oils or Fats.

Oils and fats harden with considerably less intensity
than water. The degree of hardness obtained with them is
the less the larger the cross-section of the tool, and also the
more viscid the oil or fat used. When a tool is to be given
a higher degree of hardness than is possible with oil or fat
alone, the surface of the water is coated with a layer of it
and the tool plunged through it in the water below. The
tool is thereby less rapidly cooled at the first stage, as it
has become coated with a skin of fat toughened by the heat
which retards the further cooling in the water below. The
thicker the layer of fat or oil upon the water is and the more
slowly the tool is plunged through it the less the degree of
hardness will be. For hardening cutting tools which are
to hold their edges when employed upon hard materials,
the use of water covered with oil or fat requires considerable experience and great skill if uniformly good results
are to be obtained. For such tools which are to be hardened all over, the use of milk of lime or the mode of hardening described on p. 76 is to be preferred. On the other
hand, tools which must possess a tougher degree of hardness, for instance, cutters for wood, disc knives, circular
shear knives, etc., may be hardened to advantage in water
covered with a layer of oil.

5. Oils and Fats.

Oils and fats, as previously mentioned, possess, according to their consistency, less hardening power than water. The hardness produced by them is mild with great toughness.

Thin tools, which readily crack during the process of hardening and which do not require the highest attainable degree of hardness, are hardened in oil or fat. Of the oils, petroleum hardens with the greatest intensity; next glycerine, which heretofore has not been sufficiently appreciated as a hardening fluid; then light mineral oils, and finally viscid vegetable oils, for instance, linseed oil.

Amongst fats, melted tallow and train oil are most frequently used. A somewhat higher degree of hardness is obtained in melted tallow than in oils.

In using fat or oil for hardening, it should be borne in mind that a sufficient quantity of it must be employed to allow of the article to be hardened to be vigorously moved in the bath during cooling, the same as in water, and that no rise in the temperature takes place during the operation of hardening. Too small a quantity of oil or fat is, as a rule, employed in practice, and non-success in hardening is then put down to the bad hardening qualities of the oil or fat.

6. Metals.

Mercury of all the metals possesses the greatest power of conducting heat and exerts the most energetic hardening effect. It is, however, seldom used and then only for hardening very small tools. It readily volatilizes and for hardening larger tools would have to be employed in considerable quantities. The high price of the metal and the losses in hardening by volatilization are out of all proportion to the advantages derived from its use. The vapors formed in hardening by the volatilization of mercury are of the same poisonous nature as the metal itself.

Tin, zinc and lead and their alloys in a melted state may also be used as cooling agents. However, their melting points lie so high that actual hardening in them does not take place to the same extent as in the previously-described cooling agents. The properties of steel cooled in melted metals also undergo a change, its strength being considerably increased, and it becomes so hard that it can be worked only with difficulty or not at all. It possesses, however, but little cutting power. Its elasticity is also increased.

The hardness of the steel is still further increased if the tool after having been immersed for a short time ($\frac{1}{4}$ to 2 minutes according to its cross-section) in the melted metal, is very rapidly cooled in water.

This process of hardening is suitable for springs as well as tools to be used for working soft materials, and to which the necessary tough hardness and cutting power are to be given without first tempering them after hardening, as is frequently desired in manufacturing on a large scale. Further, for giving the necessary properties to machine parts subject to great demands as regards strength and wear.

In using melted metals it is of importance that the same degree of temperature be constantly maintained. Hence adequate quantities of them must be used, proper heating devices provided, and the temperature tested with a pyrometer.

Further details, especially as regards the devices employed in the manufacture of tools on a large scale, would carry us too far.

7. Gaseous Cooling Agents.

Tools of small cross-sections may be hardened by a sharp cold current of air. However, the use of air or gases for hardening is a very limited one, and not very reliable in practice. For hardening of tools in general it is of no importance.

8. Solid Bodies as Cooling Agents.

Solid bodies of good heat-conducting power may find practical application for hardening very thin tools. Pieces of wood thoroughly saturated with water between which the thin tool is laid for the purpose of hardening are seldom used, but hardening is more frequently effected, especially in the manufacture of saws, between iron plates which are constantly cooled by an uninterrupted stream of water. The tools while between the iron plates being also subjected to pressure, this method of hardening is also termed "hardening by pressure." It is especially applicable to the continuous hardening of band-steel, and to the manufacture on a large scale of thin tools which are to acquire during the operation of hardening a tough serviceable degree of hardness.

Moist sand and clay also effect hardening, but it is difficult to obtain an even hardness by their use.

The cooling agents described above may, of course, be used in any combination desired by effecting cooling in succession in two, or more seldom in three, fluids differing in action. The process most commonly employed is to harden in water until all glow has disappeared and subsequently allowing to cool completely in oil or hot water.

By frequently repeated experiments regarding the cutting

power of medium hard tool-steel hardened in various cooling agents, the following results were obtained in Bismarckhuette:

Steel with 1 per cent. carbon.

Hardened in oil, efficiency attained.................................... 100

Hardened in tallow, efficiency attained 108

Hardened in pure water of 65° F., efficiency attained.................. 133

Hardened in water with 1 per cent. sulphuric acid, efficiency attained ... 140

From the relative proportions of these figures, a conclusion may be drawn as to the effect of the different hardening fluids used.

XIII.

TEMPERING OF HARDENED STEEL AND DEVICES FOR THIS PURPOSE.

THE main points in regard to tempering of steel have been referred to on p. 22, and the table found there.

Tempering may be effected in either one of three ways, namely:

1. By not allowing the hardened steel to cool entirely, but to effect tempering by utilizing the heat present in the interior of the tool or in a portion of it, as it progresses towards the hardened part.

2. Cooling off the tool entirely in hardening, and effecting tempering from the outside.

3. By tempering the tool from the interior and assisting the progress of the heat from the exterior.

Tempering from the interior is effected with all kinds of

tools which have been partially hardened and in which the heat stored back of, or inside, the hardened portions is allowed to progress towards them, for instance, turning knives, hand chisels, hot and cold chisels, milling tools, drills, tools for working stone, etc., or swages, cutters, hammers, etc.

To be able to recognize the advance of the heat by the progression of the temper colors, the hardened tool is rubbed bright, and should the heat advance unevenly, the respective portion is cooled by immersion for a short time in water.

When the tool shows the desired temper color, it is gradually cooled by being repeatedly plunged for a short time in water.

To impart special toughness to a tool which in use is exposed to shock and blow, it may be repeatedly tempered by rubbing off the first temper color and again producing it. For cutting tools this method may unhesitatingly be recommended especially with the use of harder steel and when the tools in use are exposed to rebounding blows, as, for instance, hot and cold chisels, which in notching rails, beams, etc., are subject to vigorous blows with heavy hammers; further, riveting punches and different tools used in structural work. Tools to be partially hardened, which have been cooled too far back of the edges, snap readily after having been tempered, in consequence of some portions of them not possessing a sufficient degree of toughness, especially if the cutting parts work under high pressure, for instance turning knives, or are subjected to shock and blow (chisels, drills, etc.).

When tools partially to be hardened have not been hardened far enough back of the edges, they possess an insufficient degree of hardness, and the operation has to be repeated. Tools which show the required degree of hardness only on the edges crack readily in use when subject to pressure and shock, the cracks running in a perpendicular direction to the edges. The material lying immediately back of the edge, and which is mostly of a smaller cross-section, is upset and the hardened edge tears in consequence of not possessing sufficient toughness to follow the change in form.

FIG. 56.

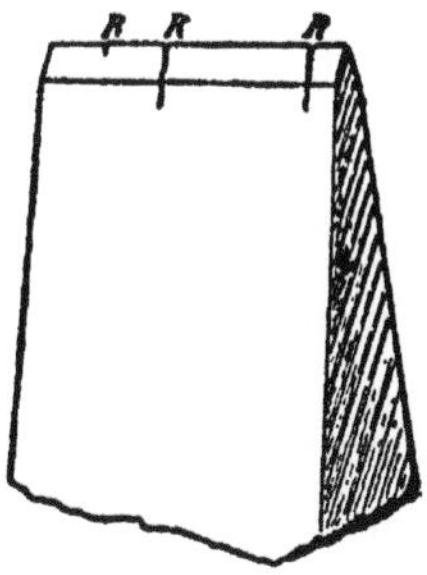

The progression of such cracks R is sketched in Fig. 56.

Tools which have been entirely hardened are seldom tempered from the interior, the process being as follows:

The tool, according to its thickness, is cooled only long enough for the surface to become cold, and so that sufficient heat remains stored in the interior which, in its progress toward the exterior, causes the previously hardened surface to be tempered.

However, since by this process the heat advancing from the interior does not, as a rule, arrive at the same time at all portions of the surface, the result will be an uneven degree of tempering, and is difficult to judge by the progression of the temper-color, because in the short space of time at the disposal of the workman it is generally impossible to brighten every portion of the hardened tool.

If tools are hardened in water and then allowed to cool in oil, tempering from the interior results very uniformly, because the surrounding oil cools off the portions where the heat may advance too energetically.

It must, however, be borne in mind that frequently the heat does not advance with sufficient rapidity from the interior to increase at the proper time the toughness of the hardened steel along the line of least strength—between the hard shell and the mild hard core. It may then happen that a severance of the hardened surface, or portions of it, results before tempering is effected ; the water in this case having cooled the tool too much previously to its having been brought into the oil.

This drawback may be obviated by assisting tempering from the interior by simultaneous heating from the outside, as described on page 76.   Cooling can then be effected to a sufficient depth to prevent softening by subsequent letting-down, and the surface thus protected from cracks.

Tools thus tempered are, however, mostly used with the full degree of hardness attained without softening by further tempering from the outside.

Attention may here be drawn to an error which is frequently committed.   In completely cooling hardened and tempered tools they should be uniformly surrounded by the cooling fluid, whether the latter be oil or water.   A mistake is frequently made in laying the hardened tool for the purpose of cooling upon the bottom of the vessel containing the cooling fluid.   On the place of contact between the side of the vessel and the tool, the latter, for reasons which can be readily understood, becomes softer than in other places. Hence the tool should be suspended in the fluid.

Tools of small cross-sections which cannot be tempered during the hardening operation are hardened entirely, and immediately after this operation are brought into hot water, or somewhat reheated in hot sand, to avoid the formation of cracks after hardening.

Tempering of hardened tools may also be effected over a moderately hot charcoal fire, upon hot sand, or in molten metals, etc.

In tempering from the outside it must be borne in mind that, by reason of their slighter thickness, projecting portions of tools, cutters, etc., may be readily heated too much and lose more hardness than originally intended. Hence, tempering should not be undertaken at a greater heat than necessary for bringing out the temper color.

A very uneven appearance of the temper color is an indication of too rapid or uneven heating. In handling tools with long edges the temper color frequently does not progress with uniformity in consequence of uneven heating. The portions which have become heated too soon are then cooled off with moist rags or by sprinkling with water until the temper color has progressed with uniformity in the other portions.

Longer tools, one end of which is to be hard while the hardness is gradually to decrease towards the outer end, are tempered by heating very slowly the end which is to be softer. The temper colors then appear at longer intervals, and, as the heat spreads, can gradually progress towards the outer end of the tool.

The tempering of tools requires above all uniform heating, just the same as in hardening. The degree of heat attained can, however, be readily judged by the progression of the temper colors, and hence great attention rather than skill is required for the operation.

The space of time in which tempering results is also of great influence upon the degree of toughness attained by the steel. The more slowly tempering is effected, the more evenly the heat will be distributed over the entire cross section of the tool, and the greater the degree of toughness will be which the tool acquires.

As will be seen from Table II. (facing p. 22), with rapid heating the yellow temper color appears when the tool has acquired a temperature of 442.4° F. By exposing the tool for a longer time to this temperature every temper color up to dark blue appears one after the other without the tool having been heated to a higher temperature. The smaller the cross section of a tool is, the more rapidly the temper colors appear.

A similar phenomenon may be observed when hardened tools are for some time laid in boiling water, their toughness being thereby materially increased while their hardness is reduced.

This phenomenon is of practical importance when tools are to be tempered in molten metals, because with the use of one and the same bath of fixed temperature different degrees of toughness and hardness can be obtained, according to the length of time the bath is allowed to act. The metal bath must of course be provided with a pyrometer, and the duration of immersion accurately regulated.

This process is practically applied in manufacturing on a large scale, especially in tempering the back portions of projectiles, etc.

Springs are in many cases tempered by what is termed blazing off with oil. The hardened spring is brushed over with oil and heated until the oil inflames and burns off. In manufacturing on a large scale this process, however, is but seldom employed, because tempering in a muffle heated to a dark-brown heat is cheaper and the result more uniform and assured.

Small tools which are to possess the hardness of a spring may be tempered in an oil which may be heated to 554° F. before it inflames.   For this operation the oil is heated in an iron vessel containing the articles to be tempered until it commences to bubble.   The vessel is then quickly covered with a close-fitting lid, removed from the fire and allowed slowly to cool.

This mode of tempering is very suitable in manufacturing on a large scale if devices are provided by means of which

Fig. 57.

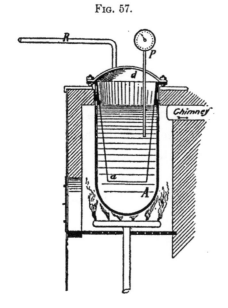

the rise in the temperature of the oil can be closely watched so that the oil is heated to near the inflaming point, but never up to it, or even above it.

The device shown in Fig. 57, which is of French origin, answers the above-mentioned requirements.

An iron kettle A is heated by a gas flame (illuminating

gas) so arranged that it can be readily turned off. A small perforated vessel a which fits in the kettle and reaches nearly to the bottom of the latter serves for the reception of the tools to be tempered. The bipartite lid d is secured by means of screw clamps. From one part of the cover, a pipe R leads to a vessel filled with water which serves for condensing the oil vapors or for carrying off oil boiling over. This part of the lid also contains the pyrometer P.

The oil is heated to about 500° F. and allowed to remain at this temperature for an accurately measured length of time. The gas flame is then turned off and the kettle allowed slowly to cool, or the tools are taken from the oil and a fresh supply of tools is introduced.

In reference to the devices for tempering it may be added that they must be adapted as much as possible to the demand for uniform heating of the tools. Even with tools only partially to be tempered, the heat must evenly extend over the portion to be tempered.

Tempering may be effected in an open fire; a bright glow or sooting flame must however be avoided, and eventually a piece of sheet iron is placed between the tool to be tempered and the fire.

For tempering smaller tools red hot iron is mostly used, it being brought into close contact with the surface to be tempered.

Perforated tools are tempered by pushing a red hot mandril into the bore.

Disc-like tools are placed between red hot iron plates of smaller circumference than the tools.

Tempering is frequently effected by means of a gas flame which, however, should not be pointed.

Tempering by means of hot sand may be effected with the use of an open fire by heating the sand upon sheet-iron and then tempering the tools in it. For tempering larger

FIG. 58.

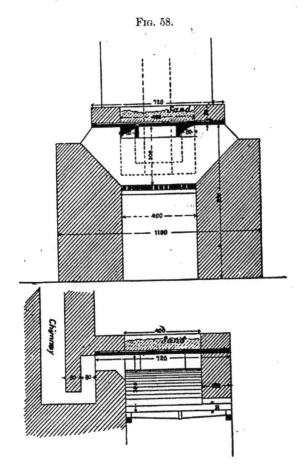

tools, or a number of them a special furnace, as shown in Fig. 58, may be used.

A perforated iron plate kept constantly red hot may be used in a similar manner. The tempering furnace may then be constructed like an ordinary kitchen hearth.

8

Partial tempering of tools in molten lead may be recommended for the purpose of producing a degree of hardness
running very evenly from one end to the other.

The portion of the tool which is to be entirely tempered
is immersed in the molten lead, which causes it to be
heated very slowly and evenly. The temper colors progress
uniformly on the portions of the tool outside of the lead.

Tempering of large tools which have to be evenly heated
in the direction of their length requires special devices to
obtain a uniform tempering temperature. A furnace which
serves for tempering long knives is shown in Fig. 59.

FIG. 59.

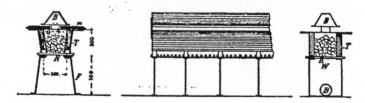

A sheet-iron box T of trapezoid cross-section rests upon
iron legs F. In place of a bottom it is furnished with iron
rods R forming a grate. The interior of the box is lined
with thin fire brick or with fire clay. The knives M to be
tempered rest upon a few cross rods o over the upper portion of the box. The whole is closed with a sheet-iron
cover B which is provided in the centre, in the direction of
its length, with an opening for the escape of the gases.

For use, glowing charcoal is evenly distributed over the
grate R and upon this is placed fresh charcoal the size of a
nut. When all the charcoal is in full glow, the knives are
placed in position as shown in the illustration and tempered.

The small furnace, above described, should be placed in a room in such a manner as to prevent the possibility of the charcoal burning unevenly by reason of draughts. If the charcoal burns less bright in some places than in others, it may be remedied by blowing with a hand bellows or by pushing pieces of sheet-iron below the grate at the places where the fire is too bright.

Fig. 60 shows the same furnace constructed of brick.

FIG. 60.

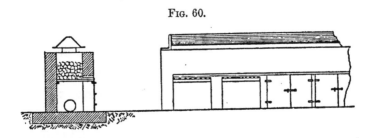

In order to be sure of the production of a uniform heat an air chamber may be fixed beneath the grate, the air being introduced by means of a perforated gas pipe (see Fig. 59). This may in a similar manner be done with the brick furnace, Fig. 60.

The temper of the steel is fixed by either quenching rapidly from the temperature attained or by allowing slowly to cool.

The tempered tool may be allowed slowly to cool if it did not contain more heat than was required for tempering.

If the tool had been partially heated, more than necessary for tempering, it must be rapidly cooled to prevent softening.

Cooling may be effected by plunging the tool in rapid succession in water or allowing it slowly to cool in mild oils or fats. If hardening of portions of the tool which are still at the hardening temperature is to be avoided, the tool for the purpose of fixing the temper is plunged in soap water.

In tempering partially hardened tools an error is frequently committed in heating them too far back of the edges, and for the purpose of cooling placing the tools— which in themselves are properly hardened and tempered— with the edges foremost in shallow water. The heat penetrating from the too highly heated portions towards the hardened parts standing in the water causes a sharp boundary to be formed between the hardened and unhardened parts, and along this boundary the edges are sure to break off. This error may be avoided by first plunging the tools in soap water to withdraw from them the highest heat. The above mentioned error is frequently made in hardening hand chisels, hot and cold chisels, drills, rock drilling tools, etc.

XIV.

STRAIGHTENING TOOLS.

Tools which have become distorted in hardening cannot be straightened while in a cold state without danger of breaking them. Hence this operation is, as far as possible, combined with that of tempering, because the steel while still in a heated state possesses sufficient plasticity to allow of being straightened.

Straightening the steel may be effected :

1. By pressure.
2. By bending and twisting by means of straightening claws.
3. By blows with the straightening and chasing hammer.
4. By uneven heating and cooling during tempering.

The distortion of tools during hardening is, in most cases, caused by uneven heating or uneven cooling, and tools of a slight cross-section and greater length or width are more frequently affected than tools of a larger cross-section. Uneven cooling causes uneven changes in volume, and consequently the steel becomes distorted.

Tools which have undergone hardening by pressure are straightened during the operation of hardening, *i. e.*, their becoming distorted is prevented during cooling.

Thin, flat tools which cannot be readily straightened with the hammer are firmly clamped, whilst in the tempering temperature, between two hand-warm iron plates and allowed to cool.

Long tools of symmetrical cross-section—for instance, twist augers, broaches, etc.—are straightened under a hand-warm screw-press.

Flat tools which become easily distorted in hardening are straightened by means of straightening-claws immediately after tempering or during that operation.

For the purpose of straightening, one end of the tool is clamped in a vice while the other end is grasped with the straightening-claw, and the tool is slowly twisted straight. In simple cases two straightening-claws may be used. Such straightening-claws are easily made by bending one

end of an iron bar, as shown in Fig. 61.   For the purpose
of straightening the tool is laid in the clefts S.

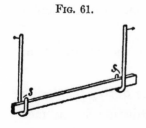

Fig. 61.

Tools distorted to an undulatory
or crescent-shaped form can be
straightened only with difficulty
by pressure, and not at all by
means of straightening claws.
Straightening must then be effected by blows, which requires
much skill and experience.

The tools required for this purpose are as follows :

A straightening anvil, *i. e.*, an anvil with a large somewhat convex face which should be smooth and thoroughly hardened.

A straightening hammer, *i. e.*, a copper hammer with
two broad faces of different convexity, and a steel
straightening hammer with a broad convex face and
a narrow sharply rounded face which stands in the
direction of the hammer-helve.

The blows with the hammer should always be given on
the hollow side of the tool.   If vigorous blows with the
flat side of the hammer do not suffice, the sharp face has to
be applied, the latter producing a powerful drawing-out
effect in the vertical direction towards the edge.   Blows
incorrectly applied readily cause worse distortion of the
tool.   The short time the latter remains in the suitable
temperature requires quick and sure work.

Straightening of tools by uneven heating and cooling in
tempering or when the tool is cold, can only be done in
rare cases and when the tool is but slightly distorted.

Such tools are laid, hollow side down, upon the tempering furnace or on a hot iron plate, and cooled upon the convex side until they have become straight.

XV.

CASE-HARDENING AND PREVENTATIVES AGAINST SUPERFICIAL DECARBONIZATION AND OVERHEATING.

THE process of case-hardening or surface-hardening is made use of if steel with but little capacity for hardening, or iron not capable of being hardened, is to be supplied with a hard surface, the object being attained by cementation.

If iron is for some time heated in intimate contact with substances rich in carbon and which readily yield it, it absorbs carbon, the latter penetrating the further from the surface towards the centre the higher the temperature is, and the longer heating has been continued. The carbon absorbed during heating imparts to the iron the capacity of being hardened, or increases that of steel capable of being but slightly hardened.

When heating with carbon is effected at a high temperature and for a longer time, the structure of the steel or iron undergoes a change, it becoming coarsely crystalline, and the cohesion between its particles is loosened. The consequence of this is a brittle shortness which is increased in the hardened state if the fine structure is not previously restored by forging. Hence to avoid failure as regards the quality of the product, the temperature should not be too high nor heating continued too long.

By case-hardening a layer of the tool extending to a greater depth is rendered harder. A tool of a smaller cross-section may even have absorbed carbon throughout and have become capable of being hardened.

It is not always desirable to impart to tools a hardness penetrating to a greater depth, a very superficial higher degree of hardness being frequently only required, and this is obtained by so-called " burning in."

The method pursued in the operation of case-hardening is as follows :

The tools are packed in an iron box with charcoal powder so as to be uniformly covered on all sides. The box is then covered with a tightly fitting lid, and after all the joints have been made air-tight by daubing with clay, it is placed in the muffle of a muffle furnace, or in an annealing furnace (see Fig. 20). Heating to a uniform temperature is effected in the same manner as in annealing, and the box is kept at this temperature for a longer or shorter time, according to the degree of hardness desired. The heat should not exceed a bright cherry-red, otherwise the tools may readily be ruined. As a rule the tools are at once hardened at this temperature and are but seldom allowed to cool and then reheated for hardening.

In practice case-hardening is resorted to in the manufacture on a large scale of tools or machine-parts which for the sake of cheapness are made of wrought iron or have been cast, for instance, for sewing machines, parts of bicycles, tools forming constituent parts of household utensils, and even scissors, knives, hatchets, etc. The carbonaceous substances employed are used either in a pure state or mixed according to their efficiency as found by experience. When charcoal is used, preference should be given to that of linden or bass-wood ; soot is less frequently employed and has no advantage over charcoal. Charcoal obtained by charring animal substances, such as leather, horn and bone, is very much liked for case-hardening, charred leather being preferred above all, since experience has proved it to be most efficacious.

Animal substances dried and ground to powder, or cut up in shavings, for instance, horn and hoof powdered or in shavings, glue, etc., are milder in their action, do not cement so rapidly as charcoal, and require longer time for heating. Hence they are but seldom used for case-hardening, but frequently as a protection of the surfaces of larger tools, to avoid superficial decarbonization or to withdraw them from the direct action of the fuel.

Tools to acquire an even degree of hardness by case-hardening must be free from adhering foreign substances,

FIG. 62.

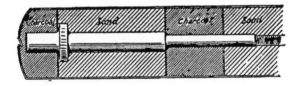

i. e., must have a pure metallic surface, and hence have to be scoured before being placed in the box.

If certain portions of the surfaces of the tools are to re-main soft, they are protected from cementation by coating them with a clay paste or packing them in substances which yield no carbon (sand, brick-dust, etc.).

Fig. 62 illustrates the manner of packing a spindle which is to be hard in two places. The spindle packed as shown in the illustration is for several hours exposed to a bright red heat, then taken from the packing and hardened.

"Burning in" requires the use of substances which yield their carbon with ease and rapidity when in contact with the tools at a higher temperature. According to experience yellow prussiate of potash is the most effective agent. When brought in contact with the red-hot tool it liquefies, and in this condition exerts a powerful cementing effect. The mode of operation is as follows:

Heat the tool to a red heat and scatter yellow prussiate of potash over the surface to be burnt in, it being best to use a fine-meshed sieve for the purpose, so as to be sure of the even distribution of the salt. The tool is then returned to the fire, heated to the hardening temperature and hardened. For a greater depth of hardness on iron or very soft steel repeat the operation twice or three times. The surface of the tool must of course be free from scale. Very small tools which are to be given a high degree of hardness are treated as follows:

Melt yellow prussiate of potash in an iron vessel over a moderate fire and bring the tool previously heated to a brown-red heat into the fused salt, and allow it to remain in it up to 15 minutes. The tool is then heated to the hardening temperature and hardened.

With small, thin tools, a similar, though somewhat milder, effect is obtained by repeatedly bringing them to a red heat, slowly plunging them in oil or fat, reheating them each time, and finally hardening in water. If the effect is to be heightened, mix with the oil or fat (train oil) sufficient soot or charcoal powder to make a pasty mass and plunge the red hot tool in the mixture. The tool becomes coated with a thick layer of the paste, which burns with difficulty and by subsequent heating effects powerful cementation.

By mixing flour, yellow prussiate of potash, saltpetre, horn shavings or hoof meal, fat and wax, a mass of paste-like consistence is obtained, which may serve for the same purpose. The different mixtures brought into commerce under the name of "hardening paste" are of similar composition, and the ingredients may be chosen at will. For instance, melt:

Wax	500 parts by weight.	
Tallow.............................	500 " "	
Rosin..............................	100 " "	

Add to the melted mass a sufficient quantity of equal parts of charred leather, horn shavings and hoof meal to make it of a paste-like consistence, then add 10 parts by weight of saltpetre, and 50 to 100 parts by weight of powdered yellow prussiate of potash, and stir thoroughly.

The tools to be burnt in are plunged, while in a red hot state, in the paste, allowed to cool in it, again heated and hardened.

A hardening powder to be scattered upon tools which are to be hardened after heating in the open fire, may be made of any desired mixture of yellow prussiate of potash, charcoal, rosin, calcined common salt, saltpetre, hoof meal, glass powder, etc. It must, however, be borne in mind that the only purpose of the substances which do not yield carbon, is to effect a better adhesion and uniform distribution of the mixture as well as to destroy oxides (scale), and therefore the carbonaceous constituents must preponderate.

The examples given below may serve as illustrations of the composition of a dry hardening powder:

Hoof meal.. 10 parts.

Charred horn .... 10 "

Saltpetre .... ½ "

Glass powder...................................... ½ "

Calcined common salt.............................. 2 "

Yellow prussiate of potash......... .... 1 "

In the manufacture of tools the agents for the production of a higher degree of hardness on the surface are of secondary importance, it being generally preferred to use steel of such hardness as will allow of the degrees of efficiency required and which can be hardened without the assistance of hardening agents.

Of greater importance, however, are agents for cooling tools which have been heated partially too rapidly or to too high a degree ; further, as a protection against contact with the combustion-gases or against partial decarbonization in heating.

Tools with sharply projecting portions on their surfaces, for instance, cutters, files, etc., can only with difficulty be heated to an even temperature without the corners and edges (teeth) becoming heated earlier, and finally to a higher degree, than the thicker body. The consequence of this is that the edges acquire too brittle a hardness, further, cracks are formed, and with long continued over-heating, decarbonization takes place and the tools remain soft after hardening.

In many cases, for instance, in hardening files, it becomes necessary to protect the teeth from overheating and decarbonization, and with other tools, for instance, cutters, to cool them during heating when they have become hot too soon.

Cooling of unevenly heated tools is effected by taking them from the fire and allowing them to cool in the air to a uniform temperature after which they are re-heated. The result of this process is, however, not always satisfactory, especially when there are considerable differences in the cross sections of the tool. In such case the use of a powder which effects cooling and prevents possible decarbonization may be decidedly recommended, and such powder should always be on hand when hardening complicated tools. A suitable powder is obtained by mixing the following ingredients:

Hoof meal	50 parts.
Rye flour	5 "
Common salt, calcined and powdered	25 "
Glass powder	½ "

Or,

Common salt, calcined	1 part.
Hoof meal	1 "
Charred leather, pulverized	1 "
Rye flour	1 "

For use scatter the powder by means of a small shovel upon the portions of the tool which at the outset show a higher heat, or dip them in the powder, repeating the operation as often as necessary. The application of the powder is of use only with complicated tools.

If the surfaces are at the outset to be protected, the tools, for the purpose of heating, are packed in a box between horn shavings and hoof meal, and heated together with the box; or the tools, previous to hardening, are coated with a firmly adhering paste of carbonaceous substances. This coat is allowed to dry and the tools are then heated as usually for hardening.

For this purpose, the mixtures given below, which are to be applied in a pasty condition, may be used :

Charcoal or charred leather .... 1 part.
Common salt .... 1 "
Rye flour .... 1 "

Or,

Hoof meal .... 4 parts.
Rye flour .... 1 part.
Yellow prussiate of potash .... 1 per cent. by volume.
Glass powder .... 1 " " ·

Or,

Hoof meal.. 2 parts.
Charred leather .... 2 "
Calcined horn meal................................ 2 "
Potassium chromate .... 1 part.
Yellow prussiate of potash........... .... 1 "
Rye flour.. 2 parts.

The ingredients, previously reduced to a fine powder and intimately mixed, are made into a paste with concentrated common salt solution. The mixture is allowed to stand quietly for a few days, then thoroughly stirred, and for use applied with a brush to the tools, which should be previously freed from adhering grease. To prevent the mass from peeling off when heated, it should be thoroughly dry before hardening the tools coated with it.

The composition of the mixtures given above may, of course, be varied in any manner as. desired and suited for the purpose in view. Only in rare cases is their use connected with any disadvantages, the principal of them being that the hardened tool shows a spotted appearance, due to the mass having in some places been burnt to it.

For case-hardening steel by cementation, carbon compounds in gaseous form are also used. If, for instance, illuminating gas is conducted over the surface of red-hot steel, it exerts a cementing effect of great energy, the surface of the steel becoming hard when rapidly cooled.

This process is practically applied to armor-plates to which a hard surface is to be given. For this purpose carburetted hydrogen, formed by the immersion of calcium carbide in water, is used. The gas is conducted over the surface of the red-hot armor-plate, the air being excluded.

Pig iron is seldom employed for surface-hardening. It contains a high percentage of carbon, which it readily yields to iron or steel when it is brought into intimate contact with it, or when it is made to fuse on the surface of red-hot iron. The practical value of this mode of surface-hardening is very small, and the same result can, with greater certainty, be attained by case-hardening, or burning in.

XVI.

WELDING OF STEEL.

THE operations in welding steel involve the overheating of the latter to a high degree, and consequently it may readily be burnt. It should be borne in mind that for welding, steel has to be heated to a less degree than iron, and only high enough to produce a pasty condition of the surface; iron, on the other hand, may be softened to a greater depth. When steel is heated to a greater depth to

a scintillating white heat, its structure loses its coherence
to such an extent that in the subsequent forging the steel
will crumble.

The formation of a soft layer on the surface of the steel is
an absolute necessity for welding. This state of softness is
obtained by bringing the steel to a bright yellow heat, the
operation being assisted by scattering upon the steel agents
readily fusible by themselves, and which protect the work
from oxidation. Such agents are: Borax, fine quartz sand,
clay, dried and pulverized or brick-dust, potash, soda, sal
ammoniac, etc., which may be used in a mixture of any
desired proportions. Borax and soda are fused before use
and pulverized after cooling.

The operation of welding is carried on as follows: The
parts to be welded are carefully scarfed and fitted together.
They are then brought to the welding heat, which for soft
steel should be dark white, for hard steel bright yellow,
and for iron scintillating white. The scale formed in heating is scraped off, and shortly before attaining the welding
heat, the welding powder is applied without, however, taking the work from the fire. The work is then quickly
taken from the fire; the parts to be welded are pressed
together, a superficial union is effected by light hammer
blows, or still better by pressure. Welding powder is
again scattered upon the weld, the work returned to the
fire and again brought to the welding heat, which, however, need not be so high as the first.

Union is now effected by more vigorous hammer blows,
and the finer structure of the steel is restored by forging
continued as long as possible. To avoid the formation of
cracks, the tool must next be allowed slowly to cool, and is
then reheated for subsequent hardening.

If the weld does not turn out satisfactorily, the respective pieces have to be removed before repeating the operation.

If, after welding, the steel shows edge cracks, it has been overheated and burnt; or it is unsuitable for welding if, after carefully repeating the operation, such cracks are still perceptible and the weld does not turn out satisfactorily.

XVII.

REGENERATION OF STEEL WHICH HAS BEEN SPOILED IN THE FIRE.

When steel has been overheated to a high degree or roasted, it is best not to work it further for tools and spend money in their manufacture, since they will be brittle and crack, or do not acquire hardness.

As a rule it is found out too late that the steel has been overheated or roasted, it being recognized only by the defects of the finished tool. Regeneration, i. e., an improvement of the spoiled steel, is then out of the question. In the manufacture of tools, nostrums for improving spoiled steel are of no appreciable value, the best means being to avoid mistakes.

Steel is *burnt* when it has been heated to such a degree as to cause its structure to be dissevered, in consequence of which the steel in a hardened state shows edge cracks. In forging such steel it crumbles; in hardening it cracks. It cannot be regenerated.

Steel is *overheated* when it has been brought to above a bright-red heat, and in consequence of it acquires a coarse

9

crystalline structure, which, according to the duration of overheating, is confined to the surface—edges and corners —or extends throughout the bulk of the steel.

If in forging tool-steel has been partially overheated, and it is considered advisable not to remove the respective portion entirely, it is allowed to cool to a cherry-red heat, and then subjected to vigorous working, which is continued to a dark-red heat. The tool having been allowed slowly to cool is then carefully reheated for subsequent hardening.

Tools which have been overheated in hardening are allowed slowly to become cold, and are then carefully reheated to the lowest hardening temperature permissible. By this process the fine structure is restored and the steel regenerated. All nostrums recommended for the regeneration of burnt or overheated steel have the same object in view, and by themselves are entirely without effect. Success in attempting to regenerate overheated steel is always doubtful.

Tool steel is *roasted* or *baked* if, with the access of air, it has for some time been exposed to a temperature which by itself is not high enough to cause overheating.

From the surface of such steel the content of carbon has been partially or entirely withdrawn. Tools manufactured from it do not acquire a sufficient degree of hardness, or remain entirely soft. If a tool is suspected of having been made of slightly roasted steel one of the previously described hardening agents, best in the form of a paste, may be used; success depends on the degree of decarbonization, and is always doubtful.

Steel is called *dead* when, while in use, it has been repeatedly heated and hardened, and in consequence has

become brittle, has lost cutting power, and cracks readily in
hardening. There are no other means of overcoming this
than by removing the worn-out edge of the tool and re-
newing it.

XVIII.

INVESTIGATION OF DEFECTS OF HARDENED TOOLS.

IF in inspecting hardened tools defects are found their
causes should be immediately determined, so as to avoid
them in subsequent operations. The art of hardening is a
very difficult one, and requires great skill and long experience if to be carried on with success. Such experience
does not alone include a superficial knowledge of the hardening process to be employed in each case, but also an
accurate perception of the causes of defects and the selection
of the methods of working by which they may be avoided.
The hardener of tools should from the start have a clear
idea as to the process by which the object in view can be
attained without probable failure, and should take measures
that the various operations in hardening can be carried on
without hindrance.

When many tools of the same kind are to be hardened,
the operation should commence with one piece, which immediately after hardening is cleaned under water by means
of a sharp brush, and then examined as to cracks, and by
means of a smooth file as to the degree of hardness. If no
defect is discovered, a second piece may be taken in hand,

the examination of which will show whether the hardening process adopted is a suitable one, and whether the operation may be continued in rapid succession. Nevertheless it must not be omitted to subject the tools immediately after hardening to a cursory examination, and subject individual pieces to a searching inspection.

With the use of too high a temperature, small tools hardened all over crack already during cooling ; this defect can at once be dealt with.

The severance of corners and edges of larger tools also occurs mostly in the hardening bath, but at a later stage of cooling, so that the tools have to be allowed to rest for some time before the hardener can be assured that the work proceeds correctly and that hardening may be continued ; otherwise it may happen that in the subsequent inspection all the hardened pieces may be found defective.

Although too great an assurance in the correct selection of the hardening process, without scrupulous inspection frequently repeated, may lead to failure, uncertainty and timidity are just as objectionable. Precaution is then carried so far that larger tools are for hours roasted in the open fire to bring them to a quite uniform heat, and nevertheless are finally hardened in an insufficient temperature. It may happen that this process is several times repeated, because in this way the tool does not acquire hardness, and the steel used is then blamed. Such uncertainty is characteristic of an inexperienced hardener.

The following hints may serve as guides in the inspection :

1. Small tools, or tools of a small cross-section, which have been partially hardened, show cracks in the

centre running in the direction of the length in case they have been too highly heated, or if cracking has been due to flaws contained in the steel before hardening. Break the thoroughly dried tool and compare the structure of the fractures with a sample of steel of the same cross-section and of the same quality, which has been carefully brought to a cherry-red heat, hardened, and then broken.

If the fracture of the tool shows a coarser structure than the sample, hardening has been effected at too high a temperature. If the structure is fine and the sides of the crack are pure and only sprinkled with hardening water which has penetrated, failure may have been caused by the use of too cold water, or to a forging strain in the tool which has not been removed by annealing. By the use of warmer water or another hardening fluid (oil or tallow), or by carefully annealing the tool, this defect may readily be overcome.

If the surfaces of the crack show a partially or entirely dark (brown-red) color, the failure is due to defective material, the dark coloration of the fracture being caused by oxidation of the steel in heating.

2. Flat tools, such as knives, etc., or tools spreading out to a broad, sharp edge, such as chisels, drills, etc., show, after hardening, curved cracks on the corners and back of the edges. This is an indication of uneven heating or over-heating of the corners and edges.

3. The corners, edges and teeth of larger tools which

have been hardened in their entirety separate during or after hardening. If, with an otherwise correct hardening process, the fracture shows a coarse structure, partial over-heating is the cause of the failure, but if the fracture shows a fine structure, failure is due to incorrect cooling.

4. Tools cracked from the interior have been incorrectly cooled if the surfaces of the crack when laid bare show a thoroughly coherent structure. If the failure has been due to a pipe or other flaws in the interior of the steel, it may be recognized on the surfaces of the crack when laid bare.

5. When cracks or separation of corners, etc., are noticed on one side of a symmetrical tool, it is indicative of uneven heating in hardening.

6. Edge cracks running vertically to the edges and showing sides of a dark and black color occur only in burnt steel.

   If the structure of the steel is coarse and of a white glistening color, the tool has been burnt in hardening; if the structure is coarse-grained but shows little lustre, the steel has been burnt in one of the operations previous to hardening.

7. If hardened tools show an uneven degree of hardness, they have been unevenly heated for hardening, or unevenly cooled by not having been sufficiently moved about in the hardening fluid, or more cooled by it on one side. Tools which have been cooled only for a short time and then allowed to become cold while in contact with the walls of the cooling vessel, are apt to be soft on the places of contact.

8. When tools show a uniform degree of insufficient hardness it is more seldom due to too slight heating, but rather to an unsuitable hardening fluid, and perhaps also to too small a quantity of the latter.

   With the use of a lower hardening temperature and in hardening larger tools, cooling fluids acting with greater intensity should be employed, and water covered with oil, pure oil or fat should be avoided.

9. Soft spots in hardened tools are formed when coal containing sulphur is used for heating. However, such spots are also formed when the hot tool is too slowly plunged in the cooling fluid, and the latter by being dashed upwards causes here and there soft spots.

10. When the surface of a hardened tool is entirely covered with a thin soft skin, which to a slight depth can readily be attacked with a file, the tool has been too slowly heated in hardening and superficially decarbonized. After removing the soft skin by grinding, the tool, as a rule, possesses a sufficient degree of hardness to permit its use, but its cutting power is of course considerably less than that of a well hardened tool.

11. Tools which have been roasted or baked and much decarbonized in any of the operations previous to hardening, do not acquire any degree of hardness whatever. The fracture of the steel shows towards the edges dark streaks of dense lustreless structure.

12. The defects due to the tools which serve for handling the steel in hardening have already been discussed on p. 86.

In case there is a doubt or uncertainty about the processes of working to be employed as regards the quality and degree of hardness of the steel to be used or the purpose for which it is to be employed, it is advisable to apply for hardening directions to the source from which the steel has been obtained.

By an immediate inspection of the hardened tool only part of the mistakes made in hardening or previous operations can be found out. The defects noticed in using the tools are generally confined to a deficiency in the degree of hardness. Such defects are generally due to the fact that the degree of hardness has been incorrectly chosen, and partially to faulty hardening. If a tool is found to possess insufficient cutting power, it may be the fault of its not being hard enough as well as of its being too hard and brittle. Microscopical particles break out of the edges of tools which have become brittle in hardening and the edges soon become dull, the impression being the same as when too soft steel becomes rapidly dull. If it is attempted, as is frequently done, to remedy this defect by still more intense hardening, the result is the reverse. To give the tool better edge-holding power, it suffices, as a rule, to remove the superficial, more brittle, layer by grinding. When the edge has to be renewed a lower hardening temperature should be tried and less intense cooling. Should the efficiency of the tool remain unsatisfactory, it may be supposed that the degree of hardness selected has been too low.

That with cutting tools of hard steel the highest attainable degree of hardness does not always yield the greatest efficiency may be seen from the experiments made at

Bismarckhuette, the results of which are given in the table below.

For these experiments several cutters were made from one and the same bar of steel, and one-half of them were carefully hardened from a bright red heat in water of 64.4° F., and without further tempering tested under the same conditions. The average efficiency of these was reduced to the figure 100. The remaining half of the cutters were heated to not so bright a red heat, hardened in water of 64.4° F., until all heat had disappeared, plunged in boiling water and allowed to cool in it.

The efficiency was better throughout, namely :

Degree of hardness of the cutters.	Intensely hardened in water.	Hardened in water, tempered in boiling water.
No. 1. Steel with 0.85 % carbon. Efficiency........	100	112
No. 2. Steel with 1.32 % carbon. Efficiency.	100	118
No. 3. Special steel with 1.57 % carbon and 4.5 % tungsten. Efficiency.	100	135

The test was made upon annealed cast-steel of the same degree of hardness as that of the cutters.

It will be seen from the increase in efficiency, up to as much as 35 per cent., that by a suitable process of cooling an appreciable gain in time and efficiency can be attained. A defect frequently observed on very hard cutting tools consists in cracks which are formed on the surface after previous grinding off of the worn-out edge and cause the uppermost layer of steel to peel off in the form of laminæ, but which may also run in various directions throughout the entire tool. Defective steel is, as a rule, blamed for these cracks, but this is generally a mistake, because they are formed by grinding upon hard, rapidly-revolving emery wheels, even if the latter run wet.

These emery wheels attack the surface of the steel, and the area pressed against them being suddenly heated to a high temperature, a change in volume takes place, in consequence of which cracks are formed. With such emery wheels rapid cooling by the water is effected after the tool has been pressed against them, but remains ineffective during that operation.

The above-described phenomenon is more frequently observed when dealing with hard cutting tools than with tools of a less degree of hardness.

In using tools with engraved surfaces, for instance, stamping tools, which work under high pressure, the sharp edges will frequently be observed to crumble. This is an indication of too high a degree of hardness, and can be remedied by laying for a short time a hot metal plate upon the tools before using them.

The corners and edges of tools which, while in use, are exposed to a high degree of heating, readily become notched in the commencement of the operation. In consequence of heating, a superficial change in the volume of the steel takes place. The steel endeavors to contract, which is prevented by the layers underneath, which are still cold, the consequence being the formation of numerous fine cracks, and the tool becomes notched. To avoid this defect, the tool should be heated before use, and the more so the higher the degree of heating to which it is exposed during the work. Tools on which slighter demands are made are heated to hand-warm, and those exposed to a higher degree of heating are heated throughout to a uniform temperature in boiling water.

The above-recommended process may be used to advantage with dies and cutters, as well as with tools which while in use are heated by blow and shock.

Incorrect, sharply-defined hardening causes the tool while in use to break off short, and this defect can then be readily noticed.

The drawbacks due to too high a demand made on the tools or by too suddenly engaging them become manifest by the destruction of the tools themselves or of their edges, and the causes of this may be readily recognized and avoided.

Without entering into a further explanation of the defects and blemishes which may originate in hardening, tempering and in using tools, it may be recommended to subject tools which appear faultless in hardening, but prove defective in use, to as careful an examination as if they had been found defective in hardening. Such examinations serve to increase the experience of the hardener.

XIX.

IMPROVING THE PROPERTIES OF STRENGTH OF STEEL.

THE object of improving the properties of strength of steel is to give it special elasticity, strength, or toughness. In most cases the production of the greatest attainable strength and toughness is aimed at.

As has been mentioned in a previous chapter, the cohesive power of the structure of steel, which determines its strength and toughness, undergoes considerable changes in the operations of hardening, letting down, and by manipulation. By the operation of annealing the strength of steel is diminished and its toughness increased. By working steel at a low temperature its strength is increased, but its toughness decreased even to brittleness.

Steel rolled or forged at a high temperature possesses less strength and greater toughness than steel worked at a low temperature. When it is of importance to give to steel for structural purposes uniform properties of strength throughout, it must be worked in a positively even temperature. Even properties of strength may be imparted to steel of varying quality by a suitable adaptation of the working temperature.

By the operation of hardening, the strength of iron which is not capable of being hardened is essentially increased and its toughness reduced. Steel, which is capable of being hardened, loses in strength and all its toughness; it becomes brittle.

By the operation of tempering, the strength and toughness of steel are up to a certain limit increased.

If letting-down is continued above the gray temper color up to the hardening temperature, the steel passes through the various stages of strength, the latter increasing to a certain tempering temperature, and then it gradually decreases until it acquires the lowest degree, when a complete annealing temperature has been attained. The toughness of the steel increases thereby constantly.

If steel is hardened and then heated to different temperatures and allowed to cool, it will possess different degrees of strength and toughness.

Similar phenomena may be observed on steel which has been cooled from the hardening temperature in fluids of different degrees of heat, for instance, in boiling water or in molten lead.

The most useful methods for improving and regulating the properties of strength of steel by a suitable combination of hardening and reheating are patented, and are practically applied in the manufacture of steel weapons, for instance, cannon, and are still more widely employed in the manufacture of machine parts on which great demands are made, railroad material, etc.

For the application of this process the construction of suitable appliances is required, as well as extensive control of the properties of strength attained by means of machines serving for testing materials.

For the manufacture of tools and their use, the methods for improving the properties of strength of steel are but of secondary importance; they are unconsciously practiced by annealing, hardening and tempering the tool.

APPENDIX.

In works in which the manufacture of tools is carried on on a large scale, the output of large quantities of products of a similar nature comes, as a rule, only into question. The appliances which serve for the various operations in forging, hardening, tempering, etc., are established; the methods of working are the result of experience and the necessity of avoiding every kind of loss, and, as a rule, have been properly selected. The same holds good as regards the skill and experience of the workmen who carry on the various operations in the manufacture of tools, and on whose attention, by reason of a suitable division of labor, but a limited demand is made.

The toolsmith or locksmith is, however, confronted by entirely different conditions. Daily, nay hourly, demands are made on him in reference to the construction of tools, as is the case in establishments where many different kinds of tools are required for carrying on the work.

In such establishments as, for instance, iron works, machine shops, etc., the rapid and uninterrupted progress of the various operations depends largely on tools of faultless construction. Varied demands are made on the foreman and workmen entrusted with this work, and to satisfy these demands, extensive knowledge and experience are required, as well as the necessary appliances. In practice such appliances are unfortunately seldom found, they being,

(142)

as a rule, so primitive as to prove rather a hindrance than an advantage in the manufacture of effective tools.

The appliances for the different operations in the manufacture of tools as far as they refer to the treatment of steel in heating and hardening have on the whole been described in the previous chapters. The process of working to be observed in these operations depends on the quality or hardness of the steel and the purpose for which the tool is to be used.

Below, the mode of working to be observed in forging, annealing, hardening, and tempering a number of tools frequently called for will be explained.

1. HAND CHISELS.

Quality of steel to be used: For locksmith's chisels for hard materials which are to hold their edges under light blows with the hammer, medium hard steel should be used; for chisels to be worked by compressed air, hard steel.

For chisels to be used upon soft materials under sharp blows of a heavy hammer, very tough steel should be used; also for chisels with long and sharply stretched-out edges.

Forging: At the utmost in a bright red heat.

Hardening: Heating to a cherry-red heat, which should extend to about 0.59 to 0.78 inch back of the edge, and then backward in gradual transition. Plunging the edge about 0.78 to 1.18 inches inches deep in water of 64° to 68° F., moving it about, then lightly up and down until all heat is quenched.

For *tempering* the chisel is rubbed bright, the progress of the temper color followed up, and after the appearance of the violet or blue color it is fixed by plunging the chisel in water, which, if required, may be repeated, or in oil or soap water.

Chisels which have to be very tough, are heated twice to the temper color, the first color being rubbed off and heating repeated.

2. Hot and Cold Chisels.

Quality of steel to be used: For hot chisels, medium hard ; for cold chisels, tough.

Forging, hardening, and *tempering,* same as for hand chisels.

If hot and cold chisels are to be subjected to particularly powerful blows, it is best to use a good quality of weldable steel to facilitate repairing of the heads, which are apt to fray and split, and to heat the edges only to a dark yellow or brown-red temper color.

3. Rivet-Chisels.

(For cutting off rivet heads.)

These are treated in the same manner as hot or cold chisels.

4. Center-Bits.

Quality of steel: Medium hard to hard.

Forging: Good cherry-red heat, if possible with the use of a charcoal fire.

Hardening: After obtaining a uniform cherry-red heat, extending to about 0.39 inch back of the edge and then gradually decreasing, thinner bits are completely cooled in water and heated to a yellow temper color by heating the tools back of the edge.

Broad, thick bits are hardened in the same manner as hand chisels and tempered from the back end with the assistance of the fire if the heat stored in the tools is not sufficient for the purpose.

5. Turning Knives and Planing Knives.

Quality of steel: Hard to very hard.

Forging: Cherry-red heat with the exclusive use of charcoal. By long continued forging, the steel readily breaks up, as well as by too powerful blows with the hammer. After forging, turning and planing knives should always be allowed to become cold.

Hardening: Slow heating, with the use of little blast, to a scant cherry-red heat, extending to about 0.78 inch back of the edge and then gradually decreasing. Quenching in water by immersing to a depth of about 1.18 inches, and moving about, and up and down. When cool, rub off and heat to a full yellow temper color and finally cool slowly by successive immersions for a short time in warm water.

6. Roll-Turning Knives

of accompanying cross-section.

Quality of steel: Very hard, special steel.

Forging: Grooved roll turning knives are cut off from profile steel or shaped from the whole steel.

Hardening: The highest attainable degree of glass hardness is demanded, which is not modified by letting-down.

Heating for hardening should not be effected in the open fire but if possible in a muffle or a hardening furnace, and should at first be done slowly to a dark cherry-red, and then rapidly to not too bright a cherry-red, heat. When the

10

latter heat has been attained, the tool is quenched until entirely cold in pure water, or better in salt water or acidulated water.

The cold knife is then laid in hot sand or hot water and allowed slowly to cool in it. With grooved turning knives hardened in their entirety, there is great danger of cracking, the latter occurring frequently, but there are no sure means of preventing it.

7. SCREW TAPS.

Quality of steel: For small taps, tough to tough hard ; for larger taps, tough hard to medium hard.

Forging: Screw taps are turned and milled from the solid piece.

Hardening: Taps worked singly are heated for hardening in an open charcoal fire by bringing the charcoal to a glow with the aid of considerable blast, then stopping the latter and bringing the tap to a dark cherry-red heat in the quiet fire thus obtained. Further heating to the hardening temperature is rapidly effected with the use of the blast.

Taps almost cylindrical in shape are plunged vertically, twisted portions foremost, in water, and very tapering taps head foremost, the latter being allowed to cool throughout, while the former must be withdrawn far enough for the head to remain soft. When the heat is quenched the taps are withdrawn from the water—the latter still evaporating on the surfaces—then let down from the interior to from yellow to brown temper color, and finally cooled entirely.

When a larger number of small taps are to be hardened at one time they are packed in a sheet-iron box between powdered charcoal and charred leather, and heated as previously described. For hardening, the taps are singly taken from the box.

In the open fire, large taps can only be brought to the hardening temperature after long-continued roasting, whereby superficial decarbonization frequently takes place and the teeth remain soft. To overcome this drawback they are heated for hardening in the hardening furnace, the teeth having been previously protected by hardening paste, or heating for hardening is effected in a muffle.

If heating in an open fire cannot be avoided, packing in a sheet-iron box between charcoal and hoof meal should be resorted to. By letting down from the interior the teeth of large pieces would be struck too late by the advancing heat, and for this reason the tools should be cooled to as great a depth as possible, then allowed slowly to cool in hot sand or hot water, so as to avoid as much as possible a severance of teeth and corners.

If letting down small taps during hardening is not desirable, they are placed immediately after cooling in hot sand or hot water. Letting down is then effected by heating the heads in a dim gas flame or spirit flame.

8. SCREW-DIES.

Quality of steel:: The same as for screw taps.

Hardening: Heating for hardening is effected in the same manner as given for screw taps. Hardening and letting down are done as follows:

The shape of screw-dies does not allow of letting down from the interior, and in hardening they must therefore be entirely cooled. To decrease the danger of cracking they should be cooled in a mild hardening agent—molten tallow—or in water until all heat has disappeared, and then allowed completely to cool in oil.

Sharply hardened screw-dies are let down at a brown-red temper color, and screw-dies of a less degree of hardness at a yellow temper color, by laying them upon a weak charcoal fire, or better upon red-hot pieces of iron.

9. BROACHES.

Quality of steel: Same as for screw taps.
Hardening: Same as for screw taps.

10. SPIRAL DRILLS FOR METAL.

Quality of steel: Medium hard to hard.

Hardening: Heating and hardening are effected in the same manner as with screw taps, but for the sake of uniform heating, in the muffle, hardening furnace or in molten lead (salts). The difficulty of hardening the entire drill lies in the fact that it must not become much or at all distorted.

In hardening from the open fire, if the use of the latter cannot be avoided, uneven heating is prevented by hardening the drill only $\frac{1}{3}$ to $\frac{1}{2}$ of its length if it is to be used for boring not very deep holes. Heating a shorter length results in an even hardening temperature, gradually decreasing backward.

Cooling is effected by plunging the drill vertically into a vessel of as great a depth as possible, moving it up and down with a rotatory motion. By moving it sideways it would readily become distorted in consequence of uneven lateral cooling.

The drill when completely cooled is heated to a yellow

temper color, and while in this state is straightened in
the screw press.

Saturated solution of common salt,
or of soda or sal ammoniac, is used as
cooling fluid.

Letting down is effected upon heated
sand or over a charcoal fire in moderate glow, which can be readily prepared as shown in Fig. 63.

Fig. 63.

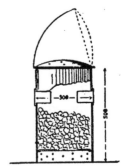

11. Cannon Drills.

Quality of steel, and hardening: Same as for spiral drills.

12. Cutters.

Quality of steel: For ordinary cutters for soft metals,
tough hard to medium hard; for cutters for hard metals
on which great demands are made, medium hard to hard;
for cutters of complicated shape for wood on which but
slight demands are made, tough to soft.

Hardening: Slot and tenon cutters are hardened in the
same manner as screw-taps and broaches.

Small cutters are heated in the muffle built in the open
fire, as shown in Fig. 3, or better in the hardening or
muffle furnace. If several cutters are to be hardened at
one time, they are packed, as described under screw-taps, in
a sheet-iron box which is heated to the hardening temperature.

The cutters are cooled singly by plunging in water and
moving them about in it; they are then placed in hot sand
and allowed completely to cool in it.

Letting down is effected by placing them upon pieces of

red-hot iron of a slighter diameter, or by means of a red-hot mandril pushed through the hole. Temper color for the bore, violet ; for the teeth, brown-red.

Large cutters of a flat shape ($\frac{1}{4}$ to about $\frac{1}{3}$ of diameter to thickness) are hardened in the same manner but only heated—supported upon pieces of sheet-iron—in the hardening furnace, but better in the muffle furnace. The teeth are protected by hardening paste, or with uneven heating, by scattering hardening powder upon them.

If permitted by the shape of the cutters, pieces of sheet-iron should be used for protection. Letting down is effected in the same manner as with small cutters.

The teeth of large or thick disc cutters, when continued on the sides, are very much exposed to being cast off in hardening if special care is not used in heating. Such cutters should not be cooled too long, but heated as soon as possible from the exterior by plunging them in hot water or covering with hot sand. Reheating to too high a degree is avoided by inspection, repeated shortly before cooling off.

Large cutters, greater in thickness than diameter, are seldom made in one piece but divided as described on p. 78. Hardening of such cutters is beset with special difficulties, and should be effected as previously described.

Profile cutters are mostly made with backward tapering teeth ; the cracking of long cutters in hardening is sought to be prevented by their suitable division.

Profile cutters with teeth which, as shown in Fig. 64, form sharp corners towards the surfaces, are protected by pieces of sheet-iron, or, as shown in the same sketch, by continuing the teeth in a curve, if permitted by the use for which the cutter is intended.

In heating, the utmost attention has to be paid to the sharp corners, and they should be frequently cooled by scattering hardening powder upon them.

Hollow cutting bodies are seldom used. They are hardened by means of a powerful jet of water, which is allowed to circulate in the hollow space.

For cooling cutters, hardening water which has been repeatedly used should be selected, or solution of common salt, or of sal ammoniac. For larger or very complicated

FIG. 64.

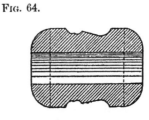

catters the hardening water is covered with a layer of oil. If, however the teeth on large cutters are of a small cross section or slight height, the use of a layer of oil should be avoided, otherwise the steel may readily remain soft.

Cutters for wood are mostly profile cutters of complicated shape, which for this reason are difficult to harden if the tools are to possess a full degree of hardness modified by subsequent letting down.

A higher degree of hardness and greater capacity of holding their edges than those of wood-working tools in general are seldom required for wood-cutters, and hence as a rule nothing more is necessary than good spring hardness.

Wood-cutters are frequently made of very hard steel and successfully used in an unhardened state. However, on account of its being more easily worked, soft steel or steel of good hardening quality is selected.

If tough steel of good hardening quality is used, it is better to harden it in a mild cooling fluid and to use it without further letting down than to give a full degree of hardness and then letting down.

Such cutters having been evenly heated to a cherry-red heat, are hardened in oil or in tallow and allowed to cool in it.

The method of cooling wood cutters of harder steel in molten metals is less known. For this purpose molten lead (741.2° F.), tin (442.4° F.), zinc (772.6° F.), or a mixture of a known fusing temperature is used, for instance,

Lead .... 8 parts.
Tin .... 4 parts.

which melts at 446° F.

The cutters heated to a uniform bright cherry-red heat are plunged, as in ordinary hardening, in the molten metal allowed for a short time to remain in it, and then rapidly cooled in water.

Cutters thus hardened possess a high degree of tough hardness, sufficient to give them the capacity of holding their edges in working wood.

If a larger number of cutters are to be cooled at one time in molten metal, a constantly even temperature is absolutely necessary for uniform success, and hence a pyrometer should be used.

13. PIPE-CUTTERS.

Quality of steel: Tough hard to medium hard.

Hardening: Inside cutters are hardened like broaches; outside cutters are hardened under a descending water-jet which strikes the bore.

14. Pipe-Cutting Knives.

Quality of steel: Tough hard to tough.

Hardening: After uniform heating to cherry-red, they are hardened in oil and used without further letting-down.

15. Milling Tools.

Quality of steel: Soft.

Hardening: Small milling tools are hardened like hot and cold chisels; large milling tools are hardened under a water jet conducted into the cavity.

Letting down is effected in the same manner as with hand and hot and cold chisels, but may be entirely omitted if a suitable quality of steel has been used.

16. Hammers.

(Hand hammers, riveting hammers, sledge hammers, etc.)

Quality of steel: As a rule, steel of a slighter degree of hardness is selected for these tools, which in hardening does not require letting-down.

Hardening: Long hammers, the faces of which can be heated separately without the heat reaching the face already hardened, are heated and hardened like hot and cold chisels, each face by itself.

Short hammers which for the purpose of hardening have to be heated in their entirety, are first hardened on the narrow face by plunging them 1.18 to 1.57 inches deep in water, then withdrawing them somewhat for allowing the hardening to spread, and finally, entirely quenching them. The broad face retains sufficient heat to be hardened in a similar manner in an ascending water-jet, the narrow face being cooled by placing wet rags upon it. When the broad face is hardened under a descending water-jet, the hardened narrow face is in the meanwhile plunged in a vessel filled with water.

While hardening both faces, the center portion of the hammer has lost sufficient heat to allow of its being cooled by repeatedly plunging for a short time in water.

17. HAMMER SWAGES.

Quality of steel: Very tough to hard; according to the purpose for which they are to be used, or to the demand made on them.

Hardening: Swages with flat faces or elevations on them are, for hardening, heated in the open forge-fire so that the face to be hardened is heated last. The swage is placed in the fire, the latter having been brought to a uniform, extensive glow by putting on a large quantity of fuel and with the use of the blast. The swage with the foot towards the blast is brought into the fire and allowed slowly to heat, the blast having been shut off enough to just maintain the glow.

When the entire tool has thus been brought to a uniform dark cherry-red heat, it is turned and the face is rapidly heated to the hardening temperature, care being taken not to overheat the corners of the swage, otherwise they may readily break off in the subsequent hardening.

During heating the scale formed on the face is scraped off and hardening powder preventing oxidation scattered upon it.

The sufficiently heated tool is then hardened by placing it, face down, upon supports in a vessel into which the water enters in a powerful ascending jet. (See Figs. 48 and 51.)

The swage is plunged 1.18 to 1.57 inches deep in water until the portion projecting from the water shows only a brown-red. It is then withdrawn to from 0.59 to 0.78 inch for the purpose of obtaining a gradual transition of the hardness, and then allowed to cool completely.

This process is subject to a modification if the swage has been made of hard, instead of mild, steel. Hardening is then interrupted after the face has been cooled, the swage is lifted from the water, and after rubbing its surface bright, is let down from the interior to a yellow to brown-red temper color; it is then further cooled in the above-described manner.

When the face of the swage is deeply engraved it cannot be hardened in ascending water, otherwise the depressions may readily remain soft. Hence, it is hardened under an ascending water jet which is uniformly distributed over the entire surface by a rose of suitable form and size, a rose fed by two supply pipes being to advantage used for the purpose. The water must, of course, enter under powerful pressure. When hardening has sufficiently progressed to allow of letting down, the supply of water is shut off, and after the face has been brought to the temper color, cooling is continued until the tool is cold.

For rapidly running hammers working with a weak blow, and for swages with very narrow faces, hard steel may be used. For swages with broad faces, especially when subject to vigorous blows, steel as tough as possible should be employed. Swages of hard steel are more subject to cracking in hardening, and also crack frequently while in use.

Defects are also frequently due to bad usage, if the

swages have been incorrectly set so that they obliquely strike one upon the other, or when thoroughly heating them throughout every time before use has been neglected.

If shortly after beginning work the surfaces of the swages show fine cracks running vertically to the edges, or criss-cross cracks, it is generally due to the fact of them not having been hardened to a sufficient depth. The steel under the hard surface is soon upset, and as the harder surface cannot follow the change in form, it cracks in different directions.

18. ROUND AND CIRCULAR SHEAR-KNIVES.

(Roll Shear-knives.)

Quality of steel : Tough.

Hardening : For hardening such shear-knives are heated exactly in the same manner as disc cutters. Cooling in the hardening fluid is effected until the knives are entirely cold, and the hardened knives are then placed in hot water or in hot sand. In letting down care must be taken that the cutting edges acquire a very uniform brown-red temper color.

19. SHEAR-KNIVES.

Quality of steel : Tough steel is used for long knives, which are principally to be used for cutting sheet metal, and for short knives for cutting iron and steel, on which great demands are made. Small shear-knives for actual cutting are made of tough hard to hard steel.

Warm shear-knives are mostly used in the unhardened state.

Hardening : Small shear-knives are best heated for this purpose in the hardening or muffle furnace. When heating in an open fire one with two or three tuyeres should be used, with charcoal as fuel.

Narrow knives are gradually brought to a throughout uniform cherry-red heat, care being taken not to overheat the corners, which otherwise in hardening break off in a curved crack.

Hardening is effected by vertically plunging in water. With the use of hard qualities of steel, holes, slots, screw-holes, etc., are filled with dry clay; with the use of tough steel this is not necessary.

Broad, short knives are heated on the edges to a uniform cherry-red heat, gradually decreasing backward. They are then hardened by slowly plunging them, edges foremost, in water, and allowed to cool until the interior heat suffices for the production of the purple-red to violet temper color, when they are allowed completely to cool in water. Such knives of very tough steel can, after thus being partially hardened, be used without letting down.

These knives may also be hardened by heating them in their entirety and cooling. Letting down is then effected by heating the backs in sand, lead, or in a quiet charcoal fire.

Great difficulties are met with in heating long shear-knives, because heating has to be very uniform in order to prevent distortion in hardening. Heating in the open fire is most difficult of all. For this purpose fires with several tuyeres, one alongside the other, are required, as well as the erection of a structure as shown in Fig. 3, so that the heat may be kept better together. Hence heating is effected to better advantage in a charcoal furnace of simple construction, such as shown in Fig. 8, in which a heat uniformly extending over the entire knife can readily be obtained.

It has to be borne in mind that the knives must first be heated from the back, and finally on the edges, so that the latter show the highest temperature before cooling.

For hardening the knives are caught on the ends and plunged backs foremost in the water and uniformly cooled.

Letting down is effected in the most simple manner, in molten lead, the knives being immersed backs foremost, or in hot sand. When neither molten lead nor hot sand is available, recourse may be had to the following device:

Set two rows of bricks, about 5¾ inches apart, upon the ground, form a grate by placing iron rods across them, and set two more rows of bricks upon the grate. The structure should be somewhat longer than the knives, and is closed at the ends by bricks. The device is shown in Fig. 65.

Charcoal brought to a uniform glow in an open fire is

FIG. 65.

put upon the grate, and the knives are placed singly, one after the other, backs foremost, in the charcoal.

Slight differences in temperature are equalized by pushing the knives back and forward during heating, and uneven letting down is prevented by cooling places which are too highly heated by means of wet rags.

20. MACHINE KNIVES FOR CUTTING PAPER, KNIVES FOR SPLITTING LEATHER, PLANING AND CUTTING KNIVES FOR WOOD.

Quality of steel : Tough to medium hard.

Hardening : The operations for hardening, heating and letting down are the same as for hardening shear-knives.

Great care must be observed in heating to prevent the sharp edges from becoming overheated, and heating must be very uniform to avoid distortion in hardening. These knives are at the best difficult to harden. They are straightened during letting down at the highest tempering temperature.

While these knives must not be too hard, but should possess great power of keeping their edges for a long time when used upon soft materials, it is preferable to harden them in tallow or train oil, the danger of warping or cracking being thereby considerably reduced.

21. STAMPING KNIVES

for stamping out of leather (soles and heels), paper, paste-board, bristol-board, etc.

Quality of steel : A good quality of weldable cast steel or weld-steel.

Hardening : For this purpose the knives are best brought in the muffle to a good cherry-red heat, the edges being previously protected with a hardening paste, and when a uniform hardening temperature has been attained, they are cooled, backs foremost, in oil or tallow.

Letting down to a yellow or violet temper color is effected by placing the knives upon their backs in hot sand, lead, or upon red-hot iron plates.

22. Circular Knives.

(Simple, straight and plate knives.)

Quality of steel: Tough to tough hard.

Hardening: Small circular knives are brought in the muffle to a uniform cherry-red heat and hardened in tallow.

Letting down is effected in the same manner as with round shear-knives. Circular knives of large diameter are brought for the purpose of hardening to a uniform cherry-red heat, care being taken to get the heat, if possible, last on the curved edges of plate knives, and to see that the heat is as uniform as possible, otherwise distortion is unavoidable.

For hardening, use a narrow long vessel with holes in the sides for fixing a crank shaft upon which the knife to be hardered is stuck. The knife should dip in the cooling fluid as far as it is to be cooled, and is then hardened by rapidly revolving it. To attain a uniform transition from the hardened edge towards the centre, water covered to a suitable depth with oil is used as cooling fluid.

23. Punches and Dies.

Quality of steel: For the punches somewhat harder steel, namely, tough hard to medium hard, than for the dies is used: for the latter, tough hard to tough.

Forging and annealing: In forging punches and dies and stamps, overheating of the working edges and corners must be carefully guarded against, it being especially injurious. The object of annealing is to facilitate the working of the steel, but to avoid partial decarbonization of the surfaces and edges these tools should be annealed packed in a box between powdered charcoal or hoof meal.

Hardening: In hardening the punches it must be borne in mind that they are almost exclusively engaged on the edges and must be capable of holding them for a long time. Hence, for hardening they are first heated from the rear, and as soon as the upper portions show a uniform dark cherry-red heat, the front portions are brought to the hardening temperature.

Hardening is effected in the same manner as with turning knives by cooling to a certain depth—as far as the punch is to be hard—and letting down from the rear of the red-hot tool to dark yellow to violet temper color. More seldom the entire punch is cooled off, and letting down effected by reheating the back portion.

The dies are mostly made of mild steel and of but a slight height. Low dies are heated for hardening to a uniform cherry-red heat, it being advisable to protect the internal cutting edges from decarbonization by coating them with hardening paste, or from overheating, by scattering hardening powder upon them. The die is then hardened in its entirety by plunging it vertically in the hardening fluid.

When made of very mild steel the die is used with its full hardness, but when made of harder steel it has to be let down, which is effected by placing it upon a red-hot iron plate until the yellow temper color appears upon the surface.

High dies are not hardened in their entirety, but by a descending water-jet which must strike with particular force the surfaces and interior openings. The process is the same as for swages.

Crumbling of the edges of dies in use, which is frequently

11

observed, is due to too high a degree of hardness; the formation of fine cracks is caused by the hardness not penetrating to a sufficient depth.

24. Stamps and Dies.

These differ from punches and dies in that they have
not a cutting effect, but transfer by pressure a more or less
sharp engraving to metals laid between them.

Quality of steel: Medium to very hard, special qualities
being particularly serviceable for this purpose.

Hardening: The duty demanded from the stamp and the
die being the same, they are made of steel of the same degree of hardness. Both are also hardened in the same
manner.

In heating stamps and dies care must be taken that the
engraved surfaces do not become decarbonized or covered
with scale. However, since these pieces have mostly large
cross-sections and must be hardened in their entirety, they
have to be heated for some time and there is always danger
of decarbonization as well as of the formation of scale.
This is prevented by heating the tools packed in a box
between horn and hoof shavings. With long-continued
heating the cementing action of charcoal or charred leather
is too great, and the use of these agents should, therefore,
be avoided.

The hardening process itself has been described on p. 76.

25. Punches and Dies for Perforating Holes in Metals.

Quality of steel: For dies, mild, tough to tough hard steel.
For punches to be used upon thin metals, tough hard to

medium hard steel; for punches for thick articles of hard metals, medium hard to hard steel, and special qualities serving for this purpose. Generally speaking, what has been said in reference to forging and hardening of punches and dies, also applies to these tools. It is, however, necessary to say something in reference to the treatment of punches on which great demands are made, for instance, in the preparation of railroad material in punching fish-plates, bed-plates, etc.

Instead of forging the working part of the punch, it is better to turn it and give it its final shape by filing. In forging, the portion of the punch on which the greatest demands are made, is readily overheated, and in upsetting, as is mostly done for the purpose of obtaining a broader surface of application, the cohesive power of the structure is injured, which, when the tool is used, results in the steel peeling off concentrically. For hardening, the punch is heated in an open fire with the exclusive use of charcoal, or better in a muffle. The thicker portion is first heated and brought to a cherry-red heat, and then the portion on which the greatest demand is made. Overheating of the edges and corners must be carefully avoided, and should they become heated too early or too highly, they are cooled by dabbing with wet rags, or by scattering hardening powder upon them.

For the purpose of hardening, the punch is plunged vertically in the hardening water so that the thick portion is also cooled to a depth of from 1.18 to 1.57 inches, the punch being moved about until cooled off. It is then withdrawn and allowed to blue from the thick portion. The blue temper color is chosen, and when the punch has been somewhat highly heated, hence has been quite sharply hardened, it is twice blued. A fine smooth file should just take hold on the tempered surface. If not cooled at all, or only in water, such punches become quite hot when used, and become covered with a firmly adhering layer of the metal which is punched. Hence cooling is effected by means of oil.

Before use the punches are heated throughout in a quiet coal fire so that they can just be touched with the hand, or, what is still better, they are placed for some time in hot, or for a shorter time in boiling, water.

26. Mandrils

for drawing metal cases and tubes.

Quality of steel: Tough hard to tough.

Hardening: Uniform heating is readily attained only in the muffle. Hardening from melted salts is especially recommended (see p. 98).

Hardening is effected by plunging the mandril vertically in water, to which a quantity of common salt may advantageously be added.

Tempering is effected by heating the head of the tool in a gas flame, alcohol flame, or by immersion in hot lead.

27. Draw Plates.

Quality of steel: Hard.

Hardening: For hardening the plates are heated in their entirety, and the inner surfaces, which are engaged in use, are protected from decarbonization by the use of hardening paste and hardening powder.

Narrow draw plates are hardened in their entirety by

cooling off in water; broad draw plates with narrow holes, by conducting a powerful water jet through the interior.

28. PILLOW-BLOCKS AND PIVOTS.

Quality of steel: Medium hard to hard.

Hardening: When large, the engaged surfaces of pivots are hardened under a falling water jet, and tempered to the pale yellow to straw-yellow temper color.

Pillow-blocks are hardened in their entirety by cooling them, apertures down, in a powerful ascending water jet. If the apertures are deep, cooling is effected under a powerful falling water jet. Subsequent letting down is effected over a faintly glowing charcoal fire, in hot sand or upon a red-hot iron plate.

29. STONE-WORKING TOOLS.

The selection of the quality of steel for these tools depends not only on the kind of tool but also on the hardness of the stone to be worked.

There is a large variety of these tools, but their construction is, as a rule, very simple, and the operations in hardening and tempering are readily carried out and require but little attention.

Since nearly all stone-working tools are subject to blow and shock and are seldom hardened in their entirety, the directions given for hardening hand chisels, hot and cold chisels, and hammers also apply to them.

When hard steel is used, the treatment must of course be very careful and overheating be avoided.

30. TOOLS FOR THE MANUFACTURE OF NAILS.

Tools serving for the manufacture of nails are subject to

various demands which have to be satisfied by selecting the suitable quality of steel and using an appropriate method of hardening.

The appliances for the operations of hardening and tempering should be selected with the greatest care, since the efficiency of the machine is dependent on the quality of the tool used and uniformly good efficiency can only be attained by uniformly good hardening.

The varied demands made on these tools and the different purposes for which they are used, do not allow of detailed directions to be given for hardening them, and the reader is therefore referred to the hardening methods and the appliances required, which have previously been described.

31. BALLS.

Quality of steel: Medium hard to hard.

Hardening: In hardening balls which chiefly serve for bearings in rapid-running machines, it should be borne in mind that they must possess surfaces as pure and glasshard as possible.

Small balls are heated, in large numbers at one time, in the muffle and allowed to drop into a deep vessel filled with water. Hardening is thus readily and completely effected without fear of the balls cracking.

On the other hand, in hardening large balls many difficulties are encountered as regards cooling off and cracking, since the strains operating from all sides towards one point may not only cause lamellar peeling off of the hard surfaces, but also cracking from the interior.

Since large balls when simply plunged in the hardening fluid do not lose, in sinking down in it, their heat with sufficient rapidity and acquire an uneven degree of hardness in contact with the walls of the vessel, it, is necessary, during cooling, to keep the hardening fluid in motion.

By handling the ball with ordinary tongs the surface of the ball would readily receive impressions and the places caught by the tongs acquire an uneven degree of hardness.

FIG. 66.

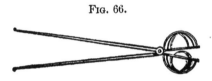

Hence tongs are used, the jaws of which are made of very thin iron in the shape of a basket, Fig. 66, by means of which the heated ball can be handled without pressure and · in hardening be brought on all sides in contact with the cooling fluid.

FIG. 67.

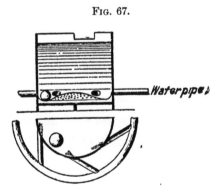

Uniform cooling may also be effected by means of the device shown in Fig. 67.

Fit a vessel, about 3.9 inches above the actual bottom, with a sieve-bottom of the shape shown in the illustration. Four or more pipes conveying in a slanting direction water under great pressure, enter the vessel 0.39 to 0.78 inch above the sieve-bottom. The upper edge of the hardening vessel is provided with a notch to allow the water to run off; and the water supply is so arranged that it can be shut off. Before commencing hardening the supply pipe is opened and the ball is thrown into the strongly agitated water, in which it sinks to the bottom and is kept in constant motion by the water flowing in.

If the hardened balls were allowed completely to cool off in the hardening bath, much waste would result from cracking. Therefore, to equalize the temperature and hardening strains, they are brought before they are entirely cooled inside into hot sand, an open fire, or hot water. It is absolutely necessary for forged balls (in contradistinction to balls turned from the solid steel) to be annealed previous to hardening in order to remove the forging strains which are formed by upsetting the corners and edges of the ball-form obtained from a cylinder or cube. Annealing is effected by one of the methods described on p. 48.

32. ROLLS.

Rolls have to be made of very hard steel, since their surfaces must possess a uniform glass hardness, and hardening them is one of the most difficult problems confronting the hardener.

The process of hardening a large roll is described by the following example:

The danger of cracking from the interior has been reduced in the construction of the roll by boring it out. The entire surface of the roll, a–a, b–b, Fig. 68, is to be hard, while the journals, z–z, are to remain as soft and tough as possible.

Previous to heating, the journals, z–z, are given a coat of loam or clay, which, to make it more binding, is mixed with cows' hair and, to prevent peeling off in consequence of shrinkage by heating, with chamotte, graphite, pulverized fire-brick, etc. Each journal is enclosed in a sheet-iron pipe of as large a diameter as the thickest part of the roll, and the mixture rammed in between the pipe and the journal. At m–m discs of sheet-iron projecting beyond the edge of the roll are arranged. Finally the bore, the ends

FIG. 68.

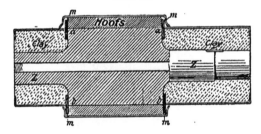

of which are provided with screw threads, are up to the latter rammed full of dry loam.

In heating the roll, which frequently requires several hours, it must be borne in mind that during this time the surface is exposed to the injurious effect of the gases of combustion, as well as to decarbonization, if suitable protection is not provided. For this purpose the roll is enclosed in a sheet-iron pipe of somewhat larger diameter, and after the space between the roll and the pipe has been rammed full of hoof shavings or soot the edges of the pipe are turned in.

The roll may now be brought into the least heated part of a reverberatory furnace of sufficient width and allowed slowly to heat. Heating in a reverberatory furnace is preferable, because other appliances, for instance, heating with charcoal, are not always available, if the roll has to be turned to attain a uniform temperature. The reverberatory furnace used should have as long a hearth as possible, so as to cause a heat gradually increasing towards the fire-place.

FIG. 69.

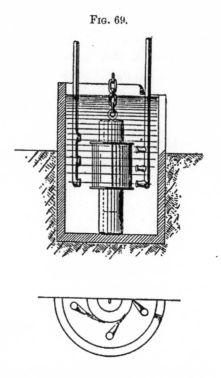

The roll is now gradually rolled into the higher heat and turned, the more frequently the hotter it becomes. When the roll is supposed to have acquired the suitable hardening temperature, a hook is screwed in the threads on the end of the journal and the roll is suspended by it by means of a chain. It is then freed from the sheet iron discs which can be readily removed, and quickly cleansed from adhering hoof shavings with a wire brush. The roll is then plunged in the hardening bath, which should be located near the furnace.

Since uniform cooling off of a roll of large diameter by moving it about in water is impossible, it is allowed to rest whilst the water is brought into vigorous motion.

This object is attained by the appliance shown in Fig. 69. It consists of a vessel of sufficient depth and of a diameter twice to four times as large as that of the roll, and is provided with pipe conduits for the introduction in a slanting direction of water under pressure.

The mouths of the pipes are made broad and slit-shaped, and a number of them are distributed at various heights so that the water around the roll is set in a vigorously whirling motion. The manner of using this appliance will be readily understood from Fig. 69.

The roll is for several hours cooled with the water flowing in, and then for some time in quiet water.

TABLE FOR THE CONVERSION OF CENTIMETERS TO INCHES.

Centi-meters.	British Inches.	Centi-meters.	British Inches.	Centi-meters.	British Inches.	Centi-meters.	British Inches.
1	0.394	51	20.079	101	39.764	151	59.450
2	0.787	52	20.473	102	40.158	152	59.844
3	1.181	53	20.866	103	40.552	153	60.237
4	1.575	54	21.260	104	40.946	154	60.631
5	1.968	55	21.654	105	41.339	155	61.025
6	2.362	56	22.048	106	41.733	156	61.418
7	2.756	57	22.441	107	42.127	157	61.812
8	3.150	58	22.835	108	42.520	158	62.206
9	3.543	59	23.229	109	42 914	159	62.600
10	3.937	60	23.622	110	43.307	160	62.993
11	4.331	61	24.016	111	43.702	161	63.387
12	4.724	62	24.410	112	44.095	162	63 781
13	5.118	63	24.804	113	44.489	163	64.174
14	5.512	64	25.197	114	44.883	164	64.568
15	5.906	65	25.591	115	45.276	165	64.962
16	6.299	66	25.985	116	45.670	166	65.355
17	6.693	67	26.378	117	46.064	167	65.749
18	7.087	68	26.772	118	46.457	168	66.143
19	7.480	69	27.166	119	46.851	169	66.537
20	7.874	70	27.599	120	47.245	170	66.930
21	8.268	71	27.953	121	47.639	171	67.324
22	8.662	72	28.347	122	48.032	172	67.718
23	9.055	73	28.741	123	48.426	173	68.111
24	9.449	74	29.134	124	48.820	174	68.505
25	9.843	75	29.528	125	49.213	175	68.899
26	10.236	76	29.922	126	49.607	176	69.293
27	10.630	77	30.315	127	50.001	177	69.686
28	11.024	78	30.709	128	50.395	178	70.080
29	11.417	79	31.103	129	50.788	179	70.474
30	11.811	80	31.497	130	51.182	180	70.867
31	12.205	81	31.890	131	51.576	181	71.261
32	12.599	82	32.284	132	51.969	182	71.655
33	12.992	83	32.678	133	52.363	183	72.048
34	13.386	84	33.071	134	52.757	184	72.442
35	13.780	85	33.465	135	53 151	185	72.836
36	14.173	86	33.859	136	53.544	186	73.230
37	14.567	87	34.252	137	53.938	187	73.623
38	14.961	88	34.646	138	54.332	188	74.017
39	15.355	89	35 040	139	54.725	189	74 411
40	15.748	90	35.434	140	55.119	190	74.804
41	16.142	91	35.827	141	55.513	191	75.198
42	16.536	92	36.221	142	55.906	192	75 592
43	16.929	93	36.615	143	56.300	193	75.986
44	17.3.3	94	37.008	144	56.694	194	76.379
45	17.717	95	37.402	145	57.088	195	76.773
46	18.110	96	37.796	146	57.481	196	77.167
47	18.504	97	38.190	147	57.875	197	77.560
48	18.898	98	38.583	148	58.269	198	77.954
49	19.292	99	38.977	149	58.662	199	78.348
50	19.685	100	39.371	150	59.056	200	78 742

INDEX.